NAVAL ANALYTICAL CAPABILITIES

IMPROVING CAPABILITIES-BASED PLANNING

Committee on Naval Analytical Capabilities and Improving Capabilities-Based Planning

Naval Studies Board

Division on Engineering and Physical Sciences

NATIONAL RESEARCH COUNCIL
OF THE NATIONAL ACADEMIES

THE NATIONAL ACADEMIES PRESS
Washington, D.C.
www.nap.edu

THE NATIONAL ACADEMIES PRESS 500 Fifth Street, N.W. Washington, DC 20001

NOTICE: The project that is the subject of this report was approved by the Governing Board of the National Research Council, whose members are drawn from the councils of the National Academy of Sciences, the National Academy of Engineering, and the Institute of Medicine. The members of the committee responsible for the report were chosen for their special competences and with regard for appropriate balance.

This study was supported by Contract No. N00014-00-G-0230, DO #20, between the National Academy of Sciences and the Department of the Navy. Any opinions, findings, conclusions, or recommendations expressed in this publication are those of the author(s) and do not necessarily reflect the views of the organizations or agencies that provided support for the project.

International Standard Book Number 0-309-09516-6

Additional copies of this report are available from:

Naval Studies Board, National Research Council, The Keck Center of the National Academies, 500 Fifth Street, N.W., Room WS904, Washington, DC 20001; and

The National Academies Press, 500 Fifth Street, N.W., Lockbox 285, Washington, DC 20055; (800) 624-6242 or (202) 334-3313 (in the Washington metropolitan area); Internet, http://www.nap.edu.

Printed in the United States of America

THE NATIONAL ACADEMIES

Advisers to the Nation on Science, Engineering, and Medicine

The **National Academy of Sciences** is a private, nonprofit, self-perpetuating society of distinguished scholars engaged in scientific and engineering research, dedicated to the furtherance of science and technology and to their use for the general welfare. Upon the authority of the charter granted to it by the Congress in 1863, the Academy has a mandate that requires it to advise the federal government on scientific and technical matters. Dr. Ralph J. Cicerone is president of the National Academy of Sciences.

The **National Academy of Engineering** was established in 1964, under the charter of the National Academy of Sciences, as a parallel organization of outstanding engineers. It is autonomous in its administration and in the selection of its members, sharing with the National Academy of Sciences the responsibility for advising the federal government. The National Academy of Engineering also sponsors engineering programs aimed at meeting national needs, encourages education and research, and recognizes the superior achievements of engineers. Dr. Wm. A. Wulf is president of the National Academy of Engineering.

The **Institute of Medicine** was established in 1970 by the National Academy of Sciences to secure the services of eminent members of appropriate professions in the examination of policy matters pertaining to the health of the public. The Institute acts under the responsibility given to the National Academy of Sciences by its congressional charter to be an adviser to the federal government and, upon its own initiative, to identify issues of medical care, research, and education. Dr. Harvey V. Fineberg is president of the Institute of Medicine.

The **National Research Council** was organized by the National Academy of Sciences in 1916 to associate the broad community of science and technology with the Academy's purposes of furthering knowledge and advising the federal government. Functioning in accordance with general policies determined by the Academy, the Council has become the principal operating agency of both the National Academy of Sciences and the National Academy of Engineering in providing services to the government, the public, and the scientific and engineering communities. The Council is administered jointly by both Academies and the Institute of Medicine. Dr. Ralph J. Cicerone and Dr. Wm. A. Wulf are chair and vice chair, respectively, of the National Research Council.

www.national-academies.org

COMMITTEE ON NAVAL ANALYTICAL CAPABILITIES AND IMPROVING CAPABILITIES-BASED PLANNING

JOHN D. CHRISTIE, LMI, *Chair*
PAUL K. DAVIS, RAND and Pardee RAND Graduate School
JOHN F. EGAN, Nashua, New Hampshire
KERRIE L. HOLLEY, IBM Global Services
SAMUEL D. KLEINMAN, Center for Naval Analyses
R. BOWEN LOFTIN, Texas A&M University at Galveston
L. DAVID MONTAGUE, Menlo Park, California
GENE H. PORTER, Nashua, New Hampshire
ANDREW P. SAGE, George Mason University
ELAINE SIMMONS, Office of the Secretary of Defense
WILLIAM D. SMITH, Fayetteville, Pennsylvania

Staff

CHARLES F. DRAPER, Director
ARUL MOZHI, Study Director
MARY G. GORDON, Information Officer
SUSAN G. CAMPBELL, Administrative Coordinator
IAN M. CAMERON, Research Associate
SIDNEY G. REED, JR., Consultant
RAYMOND S. WIDMAYER, Consultant

Staff

CHARLES F. DRAPER, Director
ARUL MOZHI, Senior Program Officer
MARY G. GORDON, Information Officer
SUSAN G. CAMPBELL, Administrative Coordinator
IAN M. CAMERON, Research Associate

Preface

The Department of Defense (DOD) has made capabilities-based planning a core concept for its future work. The Department of the Navy has made a good start at transforming its efforts to that style of planning, and in April 2004, it requested the National Research Council, under the auspices of the Naval Studies Board, to establish a committee to assess naval analytical capabilities and the Navy's capabilities-based planning. A fast response was requested. The terms of reference for the Committee on Naval Analytical Capabilities and Improving Capabilities-Based Planning are provided below. John D. Christie, senior fellow at LMI, chaired the committee. Biographical information on the committee membership and staff is presented in Appendix A.

The committee met once, from July 27 to July 29, 2004, to conduct a workshop at the J. Erik Jonsson Woods Hole Center of the National Academy of Sciences, in Woods Hole, Massachusetts. The committee heard presentations from the Department of the Navy, the Office of the Secretary of Defense, and the Joint Staff, and then outlined an initial draft report (see Appendix B for the workshop agenda).

The months between the workshop and publication of this report were spent preparing the draft manuscript, gathering additional information, reviewing and responding to external review comments, editing the report, and conducting the required security review to produce a public report.

Because only a single workshop was convened for this project, the report's findings are dependent to some degree on the expertise and background of its members. The committee attempted to reach an appropriate balance in addressing the terms of reference while not reaching conclusions based on limited data and information collected from the Navy and other elements of the Department of Defense. The objective of the report is to provide the Navy with useful, timely results and to address the most important near-term actions that the committee believes the Navy should take to improve its capabilities-based planning and analysis activities.

TERMS OF REFERENCE

At the request of the Department of the Navy, the Naval Studies Board of the National Research Council convened a workshop to assess current Department of Defense initiatives and the Department of the Navy's progress in transitioning from a requirements-based to a capabilities-based organization. The committee was also tasked to provide recommendations aimed at improving the organizational structure of the Office of the Chief of Naval Operations to best position the Chief of Naval Operations to fulfill his Title 10 (U.S. Code on Armed Forces) responsibilities. Specifically, the terms of reference for the study are as follows:

- Examine analytical processes and methods used by the Navy and other elements of the Department of Defense for capabilities-based planning. In particular, review ongoing studies (e.g., the Joint Defense Capabilities Study) and identify what has been learned to date (i.e., what works well and what has not succeeded);
- Assess current Navy capabilities-based planning processes and evaluation tools, as well as related activities being pursued within different elements of the Department of the Navy. In particular, evaluate the following: (1) the availability of appropriate data, scenarios, and analytic tools; (2) the adequacy of analytic skills and environments within the Navy; (3) the timeliness of the naval-related activities; and (4) the utility of the products;
- Address the key elements for successfully implementing capabilities-based planning. Determine methods and processes that N81 and others in the Navy can use to leverage their own (and others') activities for the benefit of the corporate Navy's integrated capabilities-based planning efforts and the interoperability of the Navy and joint staff capabilities-based planning processes; and
- Recommend a capabilities-based planning approach for the Navy based on the above considerations.

Acknowledgment of Reviewers

This report has been reviewed in draft form by individuals chosen for their diverse perspectives and technical expertise, in accordance with procedures approved by the National Research Council's (NRC's) Report Review Committee. The purpose of this independent review is to provide candid and critical comments that will assist the institution in making its published report as sound as possible and to ensure that the report meets institutional standards for objectivity, evidence, and responsiveness to the study charge. The review comments and draft manuscript remain confidential to protect the integrity of the deliberative process. We wish to thank the following individuals for their review of this report:

Larry Burns, General Motors,
Edward A. Frieman, University of California, San Diego,
Richard L. Kugler, National Defense University,
David M. Maddox, Arlington, Virginia,
John Tillson, Institute for Defense Analyses,
Raymond M. Walsh, Sr., Fairfax, Virginia, and
Larry D. Welch, Institute for Defense Analyses.

Although the reviewers listed above provided many constructive comments and suggestions, they were not asked to endorse the conclusions or recommendations, nor did they see the final draft of the report before its release. The review of this report was overseen by Gerald P. Dinneen, Lexington, Massachusetts. Appointed by the National Research Council, he was responsible for making certain that an independent examination of this report was carried out in accordance with institutional procedures and that all review comments were carefully considered. Responsibility for the final content of this report rests entirely with the authoring committee and the institution.

Contents

Executive Summary

In response to a request from the Department of the Navy, the National Research Council, under the auspices of the Naval Studies Board, established the Committee on Naval Analytical Capabilities and Improving Capabilities-Based Planning to conduct a workshop on Navy analytic methods and capabilities-based planning (CBP) activities. The committee was tasked to address key elements of CBP, examine Navy analytical processes, assess current Navy CBP processes and evaluation tools, and recommend an approach to making improvements. The study effort was to be short term, with a relatively rapid response provided to the Navy.

NAVY CHALLENGES AND OPPORTUNITY

Current U.S. defense strategy calls for the capability to deal with the full spectrum of current and future threats within the limits of available financial and personnel resources. In the past 15 years, the Department of Defense (DOD) has faced a constant stream of new challenges. Now, rather than being prepared to face a major Soviet threat and a few major regional contingencies (e.g., North Korea) in conventional warfare scenarios, the United States must be prepared both to deal with a larger number of more diverse threats with varied attributes and to do so in circumstances involving complex and uncertain risks. Many of the challenges were laid out in the 2001 Quadrennial Defense Review (QDR), which introduced the emphasis on capabilities-based planning.[1] The DOD has also

[1]Secretary of Defense Donald Rumsfeld. 2001. *Report of the Quadrennial Review*, Department of Defense, Washington, D.C., September. See also information available at <http://www.mors.org/cbp/read/AA-Korean-Presentation-Early-Spr.pdf>. Last accessed on October 18, 2004.

emphasized jointness and initiatives to transform U.S. military forces so that they may better cope with the new challenges, including threats from non-nation-states.

As follow-up to the QDR, the Secretary of Defense commissioned a task force—the Joint Defense Capabilities Study Team—led by the former Under Secretary of Defense for Acquisition, Technology, and Logistics, which recommended "a capabilities-based process for identifying needs, creating choices, developing solutions, and providing capabilities."[2] In October 2003, the Secretary of Defense accepted many of the recommendations that appeared soon thereafter in the final, published report of the task force and issued implementing guidance for adopting CBP in the budget development process.[3]

Although the basic principles of capabilities-based planning and analysis are not new, they have not been emphasized within the DOD in recent decades. Even today there is no consensus within the department about precisely what CBP is and what its essential elements are. This report gives the committee's view and seeks to contribute to the development of such a consensus. Basically, the challenge for the DOD and the Navy is, within fiscal constraints, to improve the development and execution of programs so as to ensure that they produce fully integrated joint warfighting capabilities in support of missions assigned to combatant commanders in the present or in the future. This challenge requires the management of complex risks within planned DOD funding levels to the satisfaction of the President and the Secretary of Defense.

The Navy has an opportunity to lead the Services as it improves its own internal CBP, analysis, and resource allocation efforts. The Navy can build on a historical legacy of analysis and of complex integrations across multiple functional areas, such as that of implementing layered defenses for antisubmarine and antiaircraft warfare.

WHAT IS CAPABILITIES-BASED PLANNING?

As there is no official government definition of the term *capabilities-based planning*, the committee adopted the following for purposes of this study: "Capabilities-based planning (CBP) is planning, under uncertainty, to provide capabilities suitable for a wide range of modern-day challenges and circumstances while working within an economic framework that necessitates choice."[4]

---

[2]Joint Defense Capabilities Study Team. 2004. *Joint Defense Capabilities Study: Improving DoD Strategic Planning, Resourcing, and Execution to Satisfy Joint Requirements,* Final Report, Department of Defense, Washington, D.C., January, p. iv.

[3]"Initiation of a Joint Capabilities Development Process," memorandum of October 31, 2003, from the Secretary of Defense to the Service Secretaries, Chairman of the Joint Chiefs of Staff, and others. Office of the Secretary of Defense, Washington, D.C.

[4]Paul K. Davis. 2002. *Analytic Architecture for Capabilities-Based Planning, Mission-System Analysis, and Transformation,* National Defense Research Institute, RAND, Santa Monica, Calif.,

Although concise, this definition highlights the need to deal with massive uncertainty (about both scenarios and details of assumptions in those scenarios) and to do so while considering a range of competitive options and trade-offs before making the choices necessitated by a budget. The definition given here reflects the committee's focus on future force/program planning and the resources for achieving it rather than on operational planning.[5] Capabilities-based planning may also be defined so as to include adaptive planning more explicitly.

The committee's review of the Navy's methods included looking at whether major strategic choices and trade-offs are being assessed with due regard for the element of uncertainty and whether there is concern about ensuring strategic adaptiveness to deal with currently unexpected developments. Figure ES.1 summarizes the key elements and criteria for carrying out capabilities-based planning.

NAVY-WIDE ENABLERS

This Executive Summary is organized in terms of two specific enablers of capabilities-based planning:

- *Frameworks, tools, and their use*—the conceptual framework, the analytic framework (including the building blocks of capability), and related methods and tools of analysis such as modeling and simulation; and
- *Personnel and organizations*—the resources for performing and managing key activities.

The Navy needs to address these enablers for its CBP activities both inside the Department of the Navy and within the larger DOD environment. In this report, the committee treats challenges inside the Navy separately from those for the Navy operating in the larger DOD environment, even though many of the

p. xi. *Explanatory note*: Current capabilities-based planning is very different from what became standard practice in the DOD over the past 20 years, but in some respects it harkens back to principles espoused in the 1960s (see, e.g., Charles J. Hitch and Roland N. McKean, 1965, *The Economics of Defense in the Nuclear Age*, Holiday House, New York, N.Y.). In other respects, current capabilities-based planning is different from the classic ideas of that era. For example, in the 1960s, sharp lines were drawn separating declaratory policy, programming, and operations. Today, efforts are being made to integrate them coherently. In addition, jointness was much less developed then than it is today. And, certainly, at that time analytical methods and computers did not permit either broad exploratory analysis or high-resolution detailed analysis of the sort that is possible today.

[5]See Chapter 4 for full definitions of these terms. Briefly, *future force/program planning* is the process by which the DOD builds its biennial budget proposals for the funding of future capabilities, whereas in *operational planning*, operational commanders devise and promulgate courses of action for any number of possible situations requiring the use of military capabilities currently available.

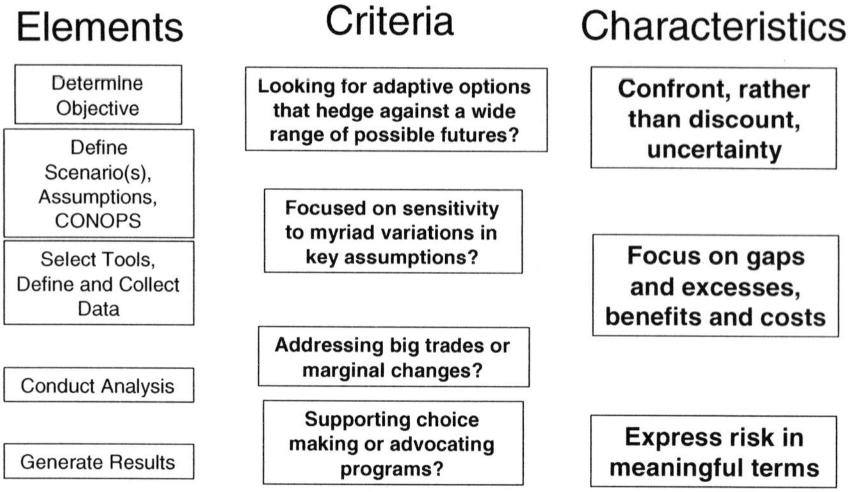

FIGURE ES.1 Key elements, criteria, and characteristics for carrying out capabilities-based planning. NOTE: CONOPS, concept of operations.

same suggestions apply in both situations. It does so because the Navy can control what it does internally but can only hope to influence the broader DOD CBP activities led by the Office of the Secretary of Defense (OSD) and the Office of the Joint Chiefs of Staff (OJCS). This latter situation is significant because the OSD and the OJCS capabilities-based planning activities are in a state of development, as are those in the Navy. Internally the Navy should carry out highly competent CBP, and in the broader DOD community it should be an integral participant, both by doing well what is required (by guidance from the Secretary of Defense) and by helping to further develop an integrated DOD process. However, the Navy should not delay in making internal improvements if the broader DOD community is slower in developing good CBP activities, is not as well focused, or cannot achieve consensus about the direction it will take.[6]

[6]Any of these eventualities could be likely in the broader DOD community, given the current lack of consensus on CBP priorities and activities across different elements of the DOD (as observed by committee members in this effort and elsewhere). A former Deputy Secretary of Defense is said to have commented that when a decision is made in industry, subordinates go about implementing it, whereas a decision in government is seen as the beginning of a dialogue. (See also the section entitled "The Navy's Current Problem and Challenge" in Chapter 1 of this report.)

Inside the Navy

Assessment of Frameworks, Tools, and Their Use

The criterion of the committee for its assessment of current Navy CBP processes, tools, and related activities is whether the Navy possesses and is using the appropriate frameworks and tools for CBP. Its summary assessment, described more fully in Chapter 3, is mixed. The committee considered issues under the categories of "conceptual framework" and "analytic framework." It considered matters at the top level of the Secretary of the Navy and the Chief of Naval Operations, at the supporting staff level of the Office of the Chief of Naval Operations (OPNAV), and at the level of system commands.

Conceptual Framework. The committee concludes that the top level of the Navy has done a creditable job in laying out a broad strategic approach under Sea Power 21. The committee is also encouraged by the evidence of managerial rethinking about organization, process, and products to support CBP at the systems command (SYSCOM) level and by the organization of a Virtual Systems Command to increase agility and integration. At the level of OPNAV studies and analysis, however, the committee finds a disconnect and a number of reasons for concern. While top-level documents and briefings emphasize jointness and the need to address a broad range of general scenarios and cases within them, the OPNAV work that the committee saw ultimately relied upon point-scenario analysis inconsistent with CBP principles. Further, it appeared that this work reflected the conceptual framework in which studies were being developed.

Analytic Framework. At the mission and operational levels, the committee saw clear examples of generic problems which often beset DOD analysis that is intended to support capabilities-based planning but actually does not. These problems include the following: trivializing of issues involving uncertainty and exploratory analysis, focusing on the scenario details preferred by the organization (e.g., those that dramatize the organization's role and needs) rather than presenting a more holistic view, and relying upon large, complex, inflexible models and databases, which often bury issues and preclude exploration.

The Navy's analysis of its aggregate-level capability needs also has problems (which exist within the broader DOD community as well).[7] The committee

---

[7]This is reflected in the "Initiation of a Joint Capabilities Development Process" memorandum of October 31, 2003, from the Secretary of Defense to the Service Secretaries, Chairman of the Joint Chiefs of Staff, and others (Office of the Secretary of Defense, Washington, D.C.), setting a goal and issuing guidance to "achieve a streamlined and collaborative, yet competitive, process that produces fully integrated joint warfighting capabilities."

is encouraged by the fact that parts of the Department of the Navy are now vigorously pursuing concepts, previously off the table, for improving aggregate effectiveness (e.g., the Fleet Response Plan and rotational crews). However, it is also concerned that such innovations are not appearing in higher-level capabilities analyses.

Based on what the committee saw and heard, it has a major concern related to the Navy's ill-developed analytic framework for portfolio-management-style analysis (i.e., strategic-level portfolio analysis for the Chief of Naval Operations (CNO) and other senior leaders).[8] The committee's view is that the analyses and presentations provided to top decision makers should have a more strategic, top-down character, and should more explicitly address economic choices and trade-offs. Certainly, they should identify shortfalls as they do now, but they should also more explicitly identify opportunities, efficiencies, and sources of funding. They should better illuminate risks and ways to mitigate or otherwise manage and balance the risks. Developing appropriate portfolio views is complex and highly dependent on the particular issue and decision context. It requires a broad spectrum of analytic capability guided by a higher level of thoughtful, strategic analysis than that typically associated with operations research or even systems analysis. This need is addressed further in the following subsection, "Assessment of Personnel and Organizations."

Recommendation 1: The Chief of Naval Operations should reiterate principles of capabilities-based planning and ensure that they are truly assimilated in Navy analytic processes.

The criteria for implementing Recommendation 1 include the following: The work accomplished should be joint and output-oriented, with the ability to actually execute operations as output. Successful CBP will require analysis over a broader scenario space, extensive exploratory analysis within selected scenarios, development of options both to solve capability problems and to achieve efficiencies, and portfolio-style assessments of those options at different levels of detail. The portfolio-style assessments should assist in making decisions on trade-offs and should address various types of risk that Navy leadership must take into

[8]The precepts of portfolio-style analysis are discussed in detail in Chapter 3. Portfolio-management analysis is the combination of hard analysis with judgment and with qualitative, value-laden trade-offs across goals—matters that are in the province of top decision makers. This combination of hard analysis with judgment may be facilitated with appropriately structured scorecards that provide holistic views of how options fare under the varying criteria that top decision makers care about. Related, spreadsheet-based tools can enable such work, provide an audit trail to assumptions, and assist in exploring the consequences of different decision-maker perspectives on the ranking of alternatives for effectiveness and cost-effectiveness. See also Richard Hillestad, Jr., and Paul K. Davis, 1998, *Resource Allocation for the New Defense Strategy: The DynaRank Decision Support System*, RAND, Santa Monica, Calif.

account. Strategic options should be adaptive, because world developments and technological developments will undoubtedly force changes, the potential need for which is not much discussed in the DOD's *Strategic Planning Guidance*.[9]

The committee is quite aware that analytic organizations have trouble responding to the demands of good capabilities-based planning. The difficulties are rooted in excessive dependence on large, complex models and related databases; in demands for detail by managers; and in the ways that analyses have been framed and conducted. Breaking these molds will not be easy. It will require a family-of-models approach and new model building. Part of this process must include "smart," low-resolution modeling and analysis (grounded in higher-resolution work or empirical data when appropriate) that puts a premium on higher-level insights rather than focusing on minutia.

Recommendation 2: The Chief of Naval Operations and the Secretary of the Navy should ensure that the Navy invests in defining and developing the new generation of analytic tools that will be needed for capabilities-based planning.

Some of the attributes needed in tools include the following: agility in low-resolution modeling coupled with the ability to go into greater depth where needed (achievable with a sophisticated family of models and games); the ability to represent network-centric operations well (including publish-subscribe architectures, rather than node-to-node representations); and the ability to deal with challenges such as those that the OSD refers to as disruptive, catastrophic, and nontraditional scenarios.

The committee is aware that the CNO has funded new work on a family of models. It is quite possible, however, that the funds will quickly be exhausted in improvements to "big models" and databases, with little benefit for higher-level capabilities-based planning, as described above. The committee encourages a balanced application of funds, including the potential purchase or use of available off-the-shelf tools.

It is not possible for the committee to make more detailed suggestions here without a more extensive study. The committee notes, however, that examples of the kinds of tools mentioned above have been developed and applied.

Assessment of Personnel and Organizations

The criterion of the committee for its assessment of current Navy CBP processes, tools, and related activities in the area of personnel and organizations

---

[9]Department of Defense. 2004. *Strategic Planning Guidance*, Secretary of Defense Donald Rumsfeld, Washington, D.C. (draft). (Classified)

is the ability of personnel and organizations to implement CBP principles. The committee's summary assessment in this area is also mixed.

The committee is encouraged by the progress made on approaches being taken in the Office of the Deputy Chief of Naval Operations (DCNO) for Manpower and Personnel (N1) and the Office of the DCNO for Fleet Readiness and Logistics (N4) with regard to personnel analyses, and by the work being done in N4 to better relate resources to readiness levels and to provide analytically based trade-off options to the CNO. The committee is also encouraged by the work presented to it by the Space and Naval Warfare Systems Command (SPAWAR), the Naval Sea Systems Command (NAVSEA), and the Naval Air Systems Command (NAVAIR), including a description of the Virtual Systems Command intended to provide integrated options and information across the previously more vertically oriented systems commands. The presentation by the Navy Warfare Development Command (NWDC) was also oriented toward some of the larger issues that need to be addressed in good capabilities-based planning, albeit with a near-term focus.

During its workshop, the committee also saw some competitive analyses work done by the Assessments Division of the Office of the DCNO for Resources, Requirements, and Assessments (N81) that related to earlier work done by the Warfare Integration Division of the Office of the DCNO for Warfare Requirements and Programs (N70). The idea and use of competitive analysis and creative tension can be productive, and the intention of generating alternative perspectives is always commendable. However, the competition between parts of the Office of the DNCO for Warfare Requirements and Programs (N6/N7) and the Office of the DNCO for Resources, Requirements, and Assessments (N8) does not appear to be helpful and involves high opportunity costs. What the committee observed during discussions with these two groups were two alternative, ad hoc views of the issue and the results of two undocumented analyses. It would be better if the Navy were to produce for the CNO an objective, well-structured analysis informed by alternative points of view. Such an analysis would be more comprehensive, systematic, parametric, questioning of assumptions, and transparent than what the committee observed.

A particularly deficient element in the Navy's current ability to support capabilities-based planning relates, as mentioned above, to providing options for choice in portfolio-style analysis modules suitable for decision making by the CNO and the Secretary of Defense.

Presenting broad, discerning, strategic-level analysis for the CNO requires a higher level of analysis than that characteristic of operations research or systems analysis. This broad analysis is in the realm of strategic planning and policy analysis. It requires a strategic sense, questioning of assumptions (even of "blessed" assumptions), applied common sense, and critical review of "how much is enough." This capability, in turn, requires having a mix of multidisciplinary warriors, policy analysts, systems analysts, engineers, economists, and managers (perhaps with master's degrees in business administration), who can effectively apply the right tools to the issues at hand. Current personnel "requirements" for OPNAV analysis are predominantly limited to capabilities and experience possessed by operations-research-oriented personnel. Those requirements need to be expanded. The committee believes that the Navy needs to change some current manpower and personnel policies to enhance its ability to build a longer-term, high-quality OPNAV staff with enhanced capabilities to perform excellent capabilities-based planning and analysis. A key element of those changes should involve creating career paths for future leaders in order to introduce such individuals early to the discipline of analytical thinking in a real-world context (e.g., the analysis for, preparation of, and review of the Navy Program Objective Memorandum (POM) and/or equivalent parts of the overall DOD program). Such career paths should continue to expose such individuals to the world of analysis and trade-offs in which outcomes influence budgets and/or major programs.

Where should such higher, strategic-level analysis be performed within the Navy? If the Chief of Naval Operations and the Secretary of the Navy (SECNAV) are to benefit fully from a first-rate analytical organization, then it is essential that the organization—

- Be able to report directly, or relatively directly, to the CNO/SECNAV (rather than being relegated to low levels in the Department of the Navy with layering to dilute its influence);
- Be institutionalized, so that it cannot easily be disbanded; and
- Be close to program builders to ensure that it reflects the reality of the issues that the decision makers must address, including funding limitations.

Whether these criteria can be met within the current OPNAV organization was not something that the committee could easily assess in the time allotted for this study. The committee has thus refrained from making specific organizational recommendations in this area. However, the committee makes the following two substantive recommendations based on the relevant government service experience of a number of its members as well as the committee's collective knowledge of the Navy, good practices in government, and good practices in industry.

Recommendation 3: The Chief of Naval Operations and the Secretary of the Navy should develop a clearly delineated concept of the Navy's future senior-level analytic support organization and define goals for its composition, including multidisciplinary orientation and officers appropriate for high positions.

The CNO and the SECNAV should insist on having analysis and presentation of decision packages that are as close to objective as possible, and on having the ready means to obtain special independent assessments on important issues when needed. In Chapter 3, the committee briefly discusses several historic models used in different Services to achieve this goal. No recommendations on a specific model would be possible without more study. In any case, the committee believes that the CNO and the SECNAV should take the time and effort necessary to ensure that they are obtaining the best analytic support possible. How they choose to organize for meeting this need could be the most important decision they make regarding the future success of the Department of the Navy's capabilities-based planning for a decade or more.

It will take considerable time to implement any decision made on this issue. Thus, the committee offers suggestions to assist the Navy for the short term.

Recommendation 4: In the short term, the Chief of Naval Operations and the Secretary of the Navy should go outside their organizations to sharpen concepts and requirements, drawing on the external community of expert practitioners in analysis. Also, they should augment their in-house analytical capabilities in the short term by drawing on Intergovernmental Personnel Act assignments (and other individuals who could take leave from their home organizations), Federally Funded Research and Development Centers, and national and other nonprofit laboratories.

Within the Larger Department of Defense Environment

Background

As discussed in Chapter 4, the committee recognizes that the DOD's capabilities-based planning is not yet fully defined. Indeed, it is both complex and confusing to participants. Nonetheless, it is the primary process that the Secretary of Defense is currently using to guide and assess Service program proposals. It is therefore important for the Navy to assume a strong role both in influencing the DOD process as it evolves and in structuring Navy program proposals in ways that are responsive to the guidance from the Secretary of Defense.

Assessment of Analytic Frameworks, Tools, and Their Use

The criterion of the committee for assessment of the use of analytic frameworks and analytic tools within the larger DOD environment is whether Navy frameworks and uses of these tools are suitable for effective work in the larger DOD environment. The summary assessment of the committee on this issue is generally negative.

The CNO's guidance to the Navy for 2004 emphasizes joint capabilities. However, in presentations at its workshop, the committee saw very little evidence of jointness in the process that the Navy uses to develop its multiyear program plan. While the Navy has done a creditable job in laying out a broad strategic approach for its own analysis and for allocation of resources, that approach is organized under the top-level components of Sea Power 21.[10] There is not yet adequate linkage between the Sea Power 21 framework and the evolving DOD capabilities-based planning framework being developed under leadership from the Office of the Secretary of Defense and the Office of the Joint Chiefs of Staff. This dichotomy is recognized by the Navy, as discussed in Chapter 4 (see Figure 4.5). However, the lack of a clear mapping from the Navy's decomposition of defense functions to the broader DOD decomposition of such functions is a potential source of a major problem that may jeopardize the Navy's credibility and effectiveness in the broader DOD capabilities-based planning and analysis processes. If this mapping problem cannot be resolved (by developing adequate crosswalks and/or by the Navy's influencing the structure of the DOD functional decomposition), the Navy may need to drop or modify its Sea Power strategic approach for its capabilities-based planning, analysis, and resource allocation efforts. It is too soon to reach a conclusion on this subject, and the Navy leadership should assess it on a periodic basis to determine if a major shift is required.

In its brief workshop, the committee also was shown no evidence of the integrated Navy and Marine Corps program planning and analyses that one would expect if the Department of the Navy was attempting to optimize its investment in such combined forces. The naval forces taken together, after all, have remarkable current and potential capabilities.

The Navy also has a long history of conducting detailed quantitative analysis at several levels of aggregation in support of both Navy program planning and the operational planning of naval component commanders. However, as mentioned above, as of mid-2004, Navy campaign analyses briefed to the committee were structured around scenarios that the Navy believes are appropriate for its purposes but that do not fully conform with the defense planning scenarios specified by guidance from the Secretary of Defense for department-wide program planning. While the Navy should certainly explore scenarios and cases that depart substantially from the DOD's baseline cases (as the guidance encourages), the credibility and effectiveness of the Navy within the DOD are jeopardized by the Navy not clearly assessing its proposed capabilities in the Secretary of Defense's mandated baselines.

On a related matter, the committee is struck by the apparent lack of Navy analysis of homeland defense (i.e., the direct defense of the United States against external attack), which is described in the Secretary of Defense's guidance as the department's highest priority. With the exception of its program to increase the capability of its Aegis ships to detect, track, and possibly engage a larger spectrum of cruise and ballistic missiles, the committee saw no evidence of Navy programs that directly address homeland defense.

---

[10]See Chapter 3 and, in particular, Figure 3.1.

Recommendation 5: The Chief of Naval Operations should direct that Navy force planning consistently include use of the baseline scenarios (including concepts of operations, threat assessments, and so on) specified by the Office of the Secretary of Defense. The Chief of Naval Operations should also work with the Secretary of the Navy and the Commandant of the Marine Corps to establish a more integrated, joint Navy/Marine Corps Program Objective Memorandum development process that could serve as a model for the Department of Defense more broadly. Navy force planning should also include extensive exploration to lead to better understanding of the consequences of scenario details and other assumptions of analysis.

Of course, within the extensive exploration recommended above, Navy planning should address in particular depth those scenarios and the cases within them that affect Navy missions, capabilities, and needs. The point here, however, is that to be a credible joint participant, the Navy should address the DOD's standard baseline cases. Navy leadership should also reassess periodically its success or failure in getting adequate linkage between Navy and DOD-wide decompositions of defense functional capabilities and adjust its program as necessary to be fully effective in the broader DOD capabilities-based planning and analysis processes.

Assessment of Personnel and Organizations

The criterion of the committee for its assessment of the effectiveness of Navy personnel and organizations in the larger DOD environment is the quality of Navy participation and influence in the OSD- and the OJCS-led capabilities-based planning efforts. The summary assessment of the committee in this area is mixed to negative.

During the workshop, Navy briefers described their organizations' attempts to cover all the bases in CBP meetings called by OSD and OJCS personnel, but indicated that the organizations had insufficient resources to participate fully and effectively. Some lack of coordination and of internal Navy scheduling control was also evident. This may be due in part to the lack of clarity in OPNAV lines of responsibility for CBP activities; that lack sometimes results in a duplication of effort. In addition, other information provided to the committee indicates that the Navy is not always well represented in the DOD joint CBP processes. Several non-Navy briefers noted that Navy representatives on the various boards and committees are often unfamiliar with the issues, not empowered to speak on behalf of the Navy, or absent. Regardless of whether these difficulties result from insufficient staff resources or other problems, they preclude justice being done regarding naval issues in the larger DOD environment and undercut Navy interests in competition for both influence and funding.

Recommendation 6: Over the longer term, the Chief of Naval Operations should identify and staff a central activity that is charged with the responsibility for harmonizing the Navy's capabilities-based planning processes with those of the larger Department of Defense (DOD) force planning community, and should staff that activity to represent Navy interests effectively in the DOD joint planning activities.

Currently, the most logical place for this assignment would be in the Office of the DCNO for Resources, Requirements, and Assessments, because the processes involved in capabilities-based planning include resource-allocation issues across all force and support areas. However, the decision on how best to accomplish the tasking for such a central activity should logically follow from or be consistent with the model chosen by the CNO to ensure having a highly competent analytical group to support the CNO in all aspects of capabilities-based planning. In the short term, to achieve the best possible utilization of resources, a possible solution would be to designate one individual (e.g., the director of the Navy staff) to be responsible for resolving any capabilities-based planning coordination problems within the OPNAV staff.

FUTURE EFFORTS

Chapter 5, "Potential Future Efforts," summarizes important sources of information that the committee was not able to investigate during this "quick-look" effort. It also identifies additional areas for further investigation that could be of benefit for the Navy to improve its capabilities-based planning.

1

Introduction

BACKGROUND

New Challenges

In military planning, resources are never sufficient to achieve all that leaders want to do with the desired minimal level of risk. In addition, over the past 15 years, the Department of Defense (DOD) has faced a constant stream of new challenges. Now, rather than being prepared to face a major Soviet threat and a few major regional contingencies (e.g., North Korea) in conventional warfare scenarios, the United States must be prepared to respond to a larger number of more diverse scenarios with varied attributes and to do so in circumstances involving complex and uncertain risks. There has always been uncertainty about threats to U.S. interests, but today, the scope of uncertainties is better recognized, making it more difficult to decide what quantities and qualities of forces are preferred within available resource limits. Decisions need to be based on assuring flexible capabilities. Changes that have contributed to this need include the following:

- In the not-too-distant past, planners often assessed the adequacy of U.S. air forces, naval forces, and ground forces separately. They performed analyses in which the Services' forces were used to a large extent independently, a reflection of reality. Over the past decade or so, however, the United States has placed great emphasis on increasing jointness, and in today's networked world, the necessity for this emphasis, including jointness at smaller unit levels within the Services, is more apparent than ever.

- In recent years, the nation's adversaries have come to recognize that they cannot deal with U.S. forces head-to-head and must therefore adopt what are sometimes referred to as the tactics of asymmetric warfare.
- And, of course, the war on terrorism has had profound and continuing effects on U.S. operations, including the extremely deep operations into Afghanistan and the increased emphasis on the need, upon occasion, to strike preventively at adversaries abroad, rather than planning only for response operations (e.g., Operation Iraqi Freedom).

As a result of these and other factors, the current world environment has created an increased demand for assessing the capabilities of packages of different types of forces with the expectation that they will need to be used in many different ways, some of them unanticipated.[1]

New Opportunities

In addition to the demand to respond to new challenges, technological advances and innovative thinking by U.S. warriors and planners provide the opportunity to develop new and different concepts of operation for using both older collections of force units and new types of force units. Thus, even if U.S. adversaries were not adaptive in responding to U.S. military operations, there would be a continuing need to assess new combinations of technology and military tactics, techniques, and procedures.

Antecedents of Capabilities-Based Planning

It is not new to the DOD to assess varied aggregations of different types of military units for use in a wide range of circumstances so that decision makers can decide how best to allocate resources among them. Nor is the concept of assessing "how much is enough" (i.e., the magnitude of capability at which decision makers believe that the risks in the given area are tolerable and that additional resources should instead be applied to a different area). As noted in a recent monograph, "capabilities-based planning is not all new, to either DOD or elsewhere. Many historical instances of related reasoning can be found, albeit sometimes by different names."[2] The same publication cites a number of ex-

---

[1]See especially Secretary of Defense Donald Rumsfeld, 2001, *Report of the Quadrennial Review*, Department of Defense, Washington, D.C., September. See also information available at <http://www.mors.org/cbp/read/AA-Korean-Presentation-Early-Spr.pdf>. Last accessed on October 18, 2004.

[2]Paul K. Davis. 2002. *Analytic Architecture for Capabilities-Based Planning, Mission-System Analysis, and Transformation*, National Defense Research Institute, RAND, Santa Monica, Calif., p. 67. For further discussion of how secretaries of defense have usually thought in terms of capabilities, see the introductory chapters of the same author's earlier book: Paul K. Davis, 1994, *New Challenges in Defense Analyses: Rethinking How Much Is Enough*, RAND, Santa Monica, Calif.

amples, including those from strategic nuclear planning in the 1960s, strategic mobility, and naval presence. Regarding strategic mobility, for example:

> The Office of the Secretary of Defense's analysis of strategic mobility issues has been particularly important over the years because funding of related systems has never been a natural organizational priority. No example of planning has ever been more obviously "capabilities based" than mobility planning, in that everyone has always understood that our strategic-mobility systems would be used in a vast range of circumstances. Scenario-based studies were . . . useful . . . [but the] scenarios were not something to be narrowly optimized against. [Looking across the scenario analysis:] It was concluded that the metric of millions of tons of miles per day was broadly useful; a sequence of studies generated reasonable goals that were then used to monitor and enforce program initiatives.[3]

Thus, in this work, scenario-based analysis was essential, but the conclusions and abstractions were more general and more robust.

While the principles of capabilities-based planning are easy to understand, they are more difficult to implement. Successful implementation requires a strategic sense, analytical thinking, applied common sense, and the questioning of assumptions (even of "blessed" assumptions). It also requires building knowledge about the particular areas of concern by small groups of personnel with a mixture of skills. Defining and analyzing the right statements of issues for analysis and decision are difficult tasks that involve much more than the development or running of large, complex, and inflexible models, which often bury issues and preclude exploration.

THE NAVY'S CURRENT PROBLEM AND CHALLENGE

The Navy needs to excel at capabilities-based planning (CBP) for at least two reasons: (1) to maximize the effectiveness of the resources that it receives and (2) to be a credible participant in the larger DOD arena, in which capabilities-based planning is the primary process that the Secretary of Defense is currently using to guide and assess Service program proposals. Moreover, as noted above, external demands on the DOD and continuing technological changes with the promise of new opportunities will require good capabilities-based planning and analysis to assist many future CNOs, secretaries of the Department of the Navy, and secretaries of the Department of Defense.

---

[3]Paul K. Davis. 2002. *Analytic Architecture for Capabilities-Based Planning, Mission-System Analysis, and Transformation,* National Research Defense Institute, RAND, Santa Monica, Calif., pp. 69-70.

The challenge for the Navy is to excel in doing its internal capabilities-based planning and analysis and to positively influence the larger DOD community in the development of the DOD-wide processes. The challenge is particularly difficult both inside the Navy and in the larger DOD environment because there is not yet a consensus within the DOD on precisely what CBP is.

Part of this lack of consensus arises from trying to integrate the former, somewhat independent decision processes for (1) overall resource allocation (programming and budgeting), (2) development and approval of "requirements" for new acquisition systems (for both hardware and software), and (3) the management of those acquisition systems. Integration will require breaking the cultural mold that many members of the Office of the Secretary of Defense (OSD), Office of the Joint Chiefs of Staff (OJCS), and military department staffs are reluctant to forgo because they believe either that the systems can operate more or less independently, or that the decision system that they themselves primarily support should be dominant in some sense. Currently, the three systems converge only at the level of the Secretary/Deputy Secretary of Defense. The Navy will have to address the same types of integration issues across its own "stovepipes" and work with the OSD and OJCS staffs as they do so department-wide.

The Navy faces many major substantive programmatic challenges that should provide its leadership with strong incentives to improve capabilities-based planning and analysis activities. Three particularly difficult challenges are as follows:

- Preparing a future Navy that can both (1) carry out joint expeditionary operations in, for example, Iraq and Afghanistan in the Middle East and in South Asia, and (2) preserve traditional strategic dominance of the seas against countries such as China that might seriously compete for it. These two strategic missions potentially require two different types of navy. Since the United States can deploy only one navy, the CBP task is to help determine the proper mix of both types of asset.
- Possibly reconsidering the vision of a larger fleet. The Navy had a vision of expanding the number of its ships to include a fixed number of different types in each category. In reality, it is not likely to possess procurement budgets big enough to fund this vision. If not, it will have to make do with fewer ships. The CBP challenge is to help determine if a smaller fleet should be based on the earlier version but with fewer ships of each type, or on a different mix of ships to be deployed (e.g., proportionately fewer submarines and more destroyers). And with CBP, the output focus needs to be on capabilities, not on numbers of ships.
- Reducing operations and maintenance (O&M) spending in order to increase funds for the procurement of new ships. An analytical challenge arises in this area because the Navy's O&M and procurement accounts cannot readily be compared. The CBP challenge is to develop metrics for both that can be compared for budgeting purposes, thereby allowing better judgments to be made

about the relationships between opportunity costs and marginal returns as resources are transferred from one account into the other.

In the time allotted for this study, the Committee on Naval Analytical Capabilities and Improving Capabilities-Based Planning did not address these challenges, nor did the Navy provide the committee with briefings about any ongoing work to use CBP to assess these challenges. Moreover, as is discussed in Chapter 3, there is no evidence that the Navy has yet developed or acquired all of the analytic tools needed to address this type of portfolio-management issue. However, with improved capabilities-based analysis, the Navy leadership could improve its resource allocation and achieve a better future Navy.

CAPABILITIES-BASED PLANNING OVER THE LONGER TERM

The briefings that this committee heard and the work that it reviewed to prepare this report in response to its tasking were focused predominantly on hardware systems and operational aspects of Navy forces. Over the longer term, for capabilities-based planning to be most effective for the Navy, and for the DOD, it must be applied to all aspects of Navy (and DOD) programs. As noted, but not emphasized, in the Joint Defense Capabilities Study (referred to as the Aldridge Study) done for the Secretary of Defense, "a large percentage of DOD's resources is devoted to enterprise operations . . . a wide range of necessary and vital support functions that enable the Department to prepare for, deploy to, execute, sustain, and rapidly recover from its military operations."[4] The study team points out that the DOD's investment in these enterprise operations and the resulting capabilities must be accounted for in the resource-allocation process, but the study does not address this large portion of the defense program in depth.

The enterprise operations referred to above are roughly the same as the activities that a report by a previous Naval Studies Board committee called the infrastructure in its broadest sense, as defined by the Joint Chiefs of Staff (i.e., all activities that provide support or control from fixed bases of operation).[5] According to that report, the Navy infrastructure accounts for about 40 percent of Navy resources, compared with about 30 percent for Navy modernization (research, development, testing, and evaluation (RDT&E) and procurement) and about 30

---

[4]Joint Defense Capabilities Study Team. 2004. *Joint Defense Capabilities Study: Improving DoD Strategic Planning, Resourcing, and Execution to Satisfy Joint Requirements,* Final Report, Department of Defense, Washington, D.C., January, pp. 3-7. The study was led by outgoing Under Secretary of Defense (Acquisition, Technology, and Logistics), E.C. "Pete" Aldridge, Jr.

[5]National Research Council, Naval Studies Board. 1998. *Recapitalizing the Navy,* National Academy Press, Washington, D.C.

percent for Navy operations (the operation of force units at sea and those that fly, including the manpower directly associated with them). Thus, it can be seen that capabilities-based planning and analysis activities limited to force operations and their modernization would exclude about 40 percent of the Navy's dollar resources.

As the Navy works over the longer term to fully implement capabilities-based planning, enterprise/infrastructure functions need to be included. It is also important to relate the inputs to these support functions to the ability of current and future combatant commanders to carry out the missions assigned to them now and in the future. If some parts of manpower, personnel, communications, and other support activities cannot be directly related to the needs of current or future combatant commanders or to the DOD's ability to implement missions assigned to it by the President, they should be candidates for possible reduction or elimination.

There are also other reasons for using capabilities-based planning and analysis to address critical support functions. For example, (1) the organization and training of individuals and force units directly affect the readiness and capabilities of the force units, and (2) manpower and personnel policies both affect the readiness and capabilities of force units (possibly at very different times). The elements in these examples also affect each other and thus can conflict if not well integrated and assessed together.

Over the longer term, the Navy (and the DOD) needs to review the resource allocation of the entire set of programs of the Navy (and the DOD) using the same capabilities-based planning and analysis principles and processes. Not to do so would risk a maldistribution of resources and the possible placement of unnecessary risk on the combatant commanders and their operational forces.

FOCUS OF THIS REPORT

Given the short duration of the committee's review, it would have been inappropriate in this report to attempt a definitive critique of the topics listed in the terms of reference for this study (see the Preface of the report). This was especially true because the Navy's efforts are evolving rapidly, and some of the impressions that the committee obtained from the review may be overtaken by events. Thus, Chapters 3 and 4 focus largely on principles and include some suggested checklists that may be of enduring value to Navy leadership. The report includes some more-detailed critique, but it does so only tentatively, to indicate where problems exist and may persist unless intervention occurs.

Consistent with this philosophy of avoiding the ephemeral, the committee also based its assessments largely on published Department of the Navy documents and on a very few key briefings presented to the committee at its July 27-29, 2004, workshop (see Appendix B). When laying out recommendations for how to proceed with capabilities-based planning, the committee drew on the philosophy clearly expressed on the subject by the Secretary of Defense, on published work cited earlier drawing on a decade's study, and on the personal experiences of its members.

ORGANIZATION OF THIS REPORT

Chapter 2 contains an overview of capabilities-based planning principles and associated risk allocation considerations. Chapter 3 presents observations and recommendations on the Navy's conceptual framework, analytic tools, and personnel and organizations and their relationship to the Navy's development of its internal capabilities-based programs to meet Navy goals. Chapter 4 offers the committee's understanding of how the Navy is participating in the ongoing joint capabilities-based processes underway in the DOD and, in particular, those involving the OSD and the OJCS. The committee also offers some recommendations as to what the Navy should do to improve its contributions to important joint defense capabilities and the related process. Chapter 5 addresses additional inputs and assessment areas that could not be covered within the committee's limited work plan and that could be of benefit to the Navy in improving its analytical activities and capabilities-based planning.

2

Key Elements of Capabilities-Based Planning and Analysis

DEFINITION OF CAPABILITIES-BASED PLANNING

There is no official government definition of capabilities-based planning, but this committee uses the following definition (which was cited in the context of the committee's tasking): "Capabilities-based planning (CBP) is planning, under uncertainty, to provide capabilities suitable for a wide range of modern-day challenges and circumstances while working within an economic framework that necessitates choice."[1]

While individual interpretations vary, capabilities-based planning (CBP) is substantially different from recent Department of Defense (DOD) "threat-based or requirements-based" analysis that was focused on point (or individual) scenarios. Indeed, the shortcoming in the earlier analysis was more the point scenarios than use of threats (obviously, planning should consider a range of specific

---

[1]Paul K. Davis. 2002. *Analytic Architecture for Capabilities-Based Planning, Mission-System Analysis, and Transformation*, National Defense Research Institute, RAND, Santa Monica, Calif., p. xi. *Explanatory note*: Current capabilities-based planning is very different from what became standard practice in the DOD over the past 20 years, but in some respects it harkens back to principles espoused in the 1960s (see, e.g., Charles J. Hitch and Roland N. McKean, 1965, *The Economics of Defense in the Nuclear Age*, Holiday House, New York, N.Y.). In other respects, current capabilities-based planning is different from the classic ideas of that era. For example, in the 1960s, sharp lines were drawn separating declaratory policy, programming, and operations. Today, efforts are being made to integrate them coherently. In addition, jointness was much less developed then than it is today. And, certainly, at that time analytical methods and computers did not permit either broad exploratory analysis or high-resolution detailed analysis of the sort that is possible today.

threats). The primary distinctions between these types of analysis are in the manner of dealing with uncertainty, in the reckoning of risk, and in the way of making choices. The core idea of the CBP approach is to confront—rather than discount—uncertainty, to express risk in meaningful terms, and to weigh costs and benefits simultaneously. The objective is to put premium value on portfolios of assets (including organizations and skill sets) that best satisfy joint needs and offer flexibility, adaptability, and robustness to hedge risk across a wide range of possible futures.

This chapter addresses the basic steps in conducting capabilities-based analysis, focusing on characteristics or success criteria, and contrasts the capabilities-based approach with the DOD's recent, more narrow, legacy methods of planning. The idea is to move away from the legacy approach, by which myriad organizations in isolation drill down on a very narrow set of scenarios with nontransparent assumptions, often for the purpose of platform or program advocacy—thus presenting top-level leadership with the problem of reconciling different results and recommendations. The goal of CBP is to harness the power of an organization to identify and analyze broad choices, provide sharp-edged implications, make clear the key assumptions and fragility of those judgments, and meaningfully express risk in the context of an unknowable future.

BASIC APPROACH FOR CAPABILITIES-BASED PLANNING

Figure 2.1 illustrates the key elements of capabilities-based planning. The CBP process starts with a formulation of what must be accomplished in order to meet strategic objectives and then proceeds to the development of a range of solutions to meet those objectives. Next, potential solutions are evaluated in a broadly framed security environment using multiple scenarios and, perhaps more importantly, parametric exploration of numerous cases within each scenario. In confronting uncertainty and vigorously exploring risk-mitigation options, CBP analysts should be searching for assets that flexibly satisfy needs across a range of situations.

In comparison, the legacy approach of point scenarios often had the effect, in Service analyses, of doing the following:

1. Protecting force structure and end-strength and replacing major platforms and systems rather than addressing demands associated with the DOD's strategic guidance and objectives;
2. Vigorously advocating a particular programmatic solution (typically platform- or system-based) instead of presenting leadership with a range of viable options and their associated cost and benefit implications; and
3. Evaluating alternatives on the basis of a single threat or a small number of "most stressing" threats with modest, single-point variations off the baseline (often employing the assumption that "less stressing" equates with less impor-

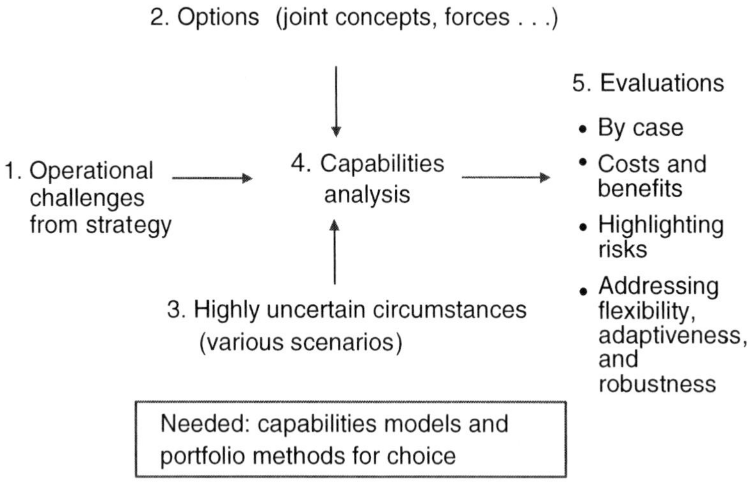

FIGURE 2.1 Key ideas in capabilities-based planning.

tant), rather than addressing various risk implications of potential solutions across a wide range of possible circumstances.

Instead of the effects of the legacy approach listed above, the following is needed:

1. Planning should focus on developing and acquiring needed capabilities, with needs for end-strength and platforms emerging as conclusions of analysis;
2. In choosing how to achieve needed capabilities, creative options should be constructed that exploit joint opportunities, new technology, and a competition of concepts; and
3. Options should be evaluated across a broad range of scenarios and circumstances so as to assure that the resulting capabilities are flexible, adaptive, and robust.

In the sections below, the key ideas in capabilities-based planning are discussed in more detail.

Operational Challenges from the Strategy

Capabilities-based analysis is conducted to provide well-articulated choices to leadership. The choices need to be distinct, viable, and bounded by economic constraints.[2] These choices should be geared to satisfying the demands of the national and the DOD strategy and associated joint needs.

Assessing joint needs involves initial direction from top-level decision makers on leadership priorities, objectives, mission assignments to combatant commanders, and risk tolerance. However, this direction on priorities and objectives should be supported by in-depth analysis that assesses capabilities, needs, and improvement options in the various mission areas essential to meeting the decision maker's objectives. With such analysis, areas of weakness or need can be identified, and priorities and objectives can be refined and even dramatically changed as analytic insights dictate. In the DOD, the proper role of the Services and agencies is not to drive demand, but to provide innovative, competitive solutions to address leadership priorities and objectives—and, as noted above, to inform changes in leadership priorities as appropriate.

In principle, the Secretary of Defense should provide broad strategic planning guidance sufficiently early in the biennial planning/programming/budgeting cycle to help frame the construction of a set of capabilities that would balance or distribute risk consistent with his objectives and priorities. In the absence of analysis to identify DOD needs, the Service leadership should engage with the staff to identify outcome-oriented capabilities (what must be accomplished) keyed to strategic objectives. This process involves comprehensively and somewhat exhaustively considering the potential operational challenges that the U.S. military may face in the future. The key is to address an adequate range of potential challenges—both from adversaries who "think like us" and, even more importantly, from those who do not.

In the formulation of these analyses to assess capabilities and identify needs, jointness should be a foundational concept from the start; it should not need to be "forced in" late in the process by the DOD leadership. The needs and capabilities analysis should consider other Services' capabilities and explore trade-offs across Service assets—admittedly a difficult task for individual Services to accomplish in isolation. But even the most basic level of capabilities-based planning and analysis implies an ability of each Service to integrate its internal analytic processes with those in joint organizations. At a minimum, this would involve using joint-approved planning scenarios as a baseline for Service analyses, both as a common starting point and as an effective way to focus on parametric exploration rather than on the creation of a Service-specific base case (discussed below). Ideally, Service analyses should contribute substantially to the corporate (DOD)

[2]A long-standing debate exists between those who claim that strategy should come first and those who claim that the budget comes first. Realistically, in good planning the two are always established together: one cannot do everything, regardless of how unconstrained a budget may be, but one also cannot have satisfactory capabilities without an adequate budget.

identification and formulation of joint needs and the competitive evaluation of solutions to those needs.

Needs formulation at the Service level can also be derived from combatant commanders' integrated priorities lists, joint experimentation, and, to some extent, joint-led studies. The value of joint-led studies has, in the past, been diminished by a consensus-based approach, a focus on a carefully scoped and scripted baseline, and an avoidance of evaluating areas of excess, or at least sufficiency.

Initially the idea is not to capture each need exactly right, but to consider carefully the implications of our current strategy as well as the impact of a change in that strategy.

This vision of defining strategic-level needs contrasts significantly with the legacy tendency to equate needs with the continued replacement of platforms or systems. In the legacy approach of recent years, up-front leadership guidance tended to be fiscally unrealistic, vague, too late in the cycle, and insufficiently focused on areas in which the organization chould tolerate more, rather than less, risk. At the working level, corporate guidance was typically referenced on briefing slides but essentially ignored in the formulation of options. In addition, in some past planning studies, needs and requirements were framed or adapted to support solutions geared to independent (stovepipe) interests.

Options in the Context of Uncertainty

Once needs are formulated, the next step is to evaluate potential solutions in a security context that acknowledges and attempts to cope with the inherent unpredictability of the future. To deal with uncertainty, the capabilities-based planning methodology calls for using multiple, fundamentally different scenarios[3] and examining myriad cases within each scenario—a technique referred to as parametric exploration.[4]

One of the many reasons for considering a wide range of circumstances and assumptions is that the baseline assumptions used in planning are often an ex-

[3]In the transformation efforts of the DOD, significant emphasis has been placed on needing analysis to support four different types of threats: irregular, catastrophic, disruptive, and traditional. The examples of capabilities-based planning in this report tend to emphasize broadening the range of scenarios and assumptions, although they may not cover the full spectrum envisioned by some leaders of DOD transformation efforts. In other instances, CBP depends less on "scenarios" than on a more straightforward, engineering-like analysis characterized by parameterization of the major variables. An example of this comes naturally in characterizing capabilities for ballistic missile defense, among other missions.

[4]For a more detailed discussion of parametric exploration and other concepts addressed in this chapter, see Paul K. Davis, 2002, *Analytic Architecture for Capabilities-Based Planning, Mission-System Analysis, and Transformation,* National Defense Research Institute, RAND, Santa Monica, Calif.

pression of system specifications, which may or may not be realizable, rather than a best estimate. This problem is common in realms in which adequate realistic testing is not feasible. In some domains, however, the Navy or the joint community has extensive databases of observed performance in exercises that provide insights into real capabilities and deficiencies that need to be addressed. These insights should inform the parametric exploration that is at the heart of CBP.

The parametric exploration approach dealing with multiple types of uncertainty is much different from that typically used during the Cold War. During that time, the Soviet Union and a few major regional threats (e.g., North Korea) clearly presented the most stressing adversaries. In addition, it was assumed that conflicts would be depicted in conventional warfare scenarios, and there was a tendency for planners to assume that they knew far more than they did about the circumstances of a conflict if it should occur. The result was not only a Soviet and major regional contingency focus, but also a focus on stereotyped wars. Smaller threats typically were ones thought to be less important that could be handled with the forces and weapons developed to counter the Soviets or major regional contingency threats. Other-than-usual versions of the Soviet scenarios were often just not given adequate attention. Threats were largely viewed as symmetrical, and lent themselves to force-on-force modeling. Force-on-force modeling rewarded incremental, "cost-effective" improvements in range, lethality, probability of kill, and force structure.

Today, adversaries are choosing strategies that do not play to our strengths and that may not even be countered effectively by conventional tactics and forces (as reflected in the increased use of special forces and the importance of international police cooperation in counterterrorism). The global war on terror has produced an adversary that seeks to attack the economic or political element of national power with a force that has a diffuse command structure, no headquarters, no territory, no tanks, and no combat aircraft. And, of course, as of September 2004, the United States was dealing with insurgencies in Afghanistan and Iraq.

Overall, then, the range of threats is now considerable. Better methods and tools are needed for analyzing diverse circumstances both within traditional conflicts and in the range of nontraditional conflicts. Proven analytic tools for assessing potential responses to these emerging nontraditional threats are scarce at best.

The challenge is how to "cover the space" (i.e., address all of the scenarios) with the desired parametric analysis. As discussed in Chapter 3, if the analysis is based on running large, complex models, prospects for adequately covering the space are not good. A wiser approach is to use a combination of low-resolution and high-resolution models, and games as appropriate, to cover the space adequately. This approach permits gathering significantly more information on the importance of different assumptions and variables than is possible by running a few cases on large, complex models. To be sure, some regions of the space will demand different concepts of operation. Developing these concepts of operation requires help from expert warfighters. However, most of the parametric work can be accomplished by analysts using relatively low resolution models, with in-depth analysis in a few places as needed.

The Evaluation Process

As shown in Figure 2.2, capabilities-based analysis consists of both policy-level analysis (big-picture reasoning, macro-level trade-offs under uncertainty) and increasingly detailed analysis such as that from systems engineering or engagement-level modeling. A hierarchy of tools and analytic techniques is needed to support the analysis, although the analysis extends well beyond the use of models. The challenge is to synthesize results from detailed analyses and extensive parametric exploration into viable trade-off options and insightful implications for resource-allocation decision makers. This process involves extensive, postmodel analytic processing—a relative rarity in deadline-driven analytic shops today.

This analytical evaluation approach involves being fluent in the details of Navy capabilities and also viewing Navy assets in the context of joint/DOD needs and programs. CBP means addressing DOD-wide (not just stovepipe-level)

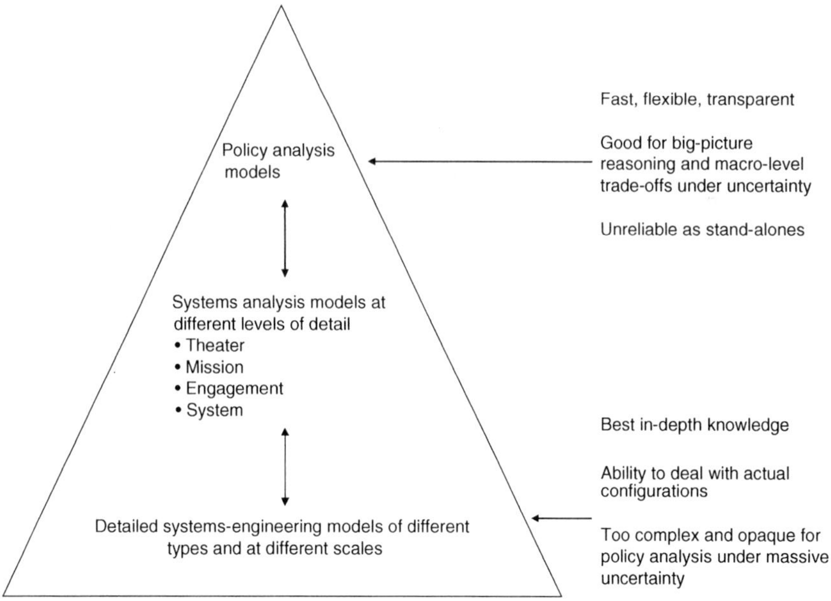

FIGURE 2.2 Family-of-models approach.

excesses as well as gaps and looking at benefits and costs of the various alternatives simultaneously. Without the ability to accept the need to distribute risks across different areas and thereby reduce capabilities in some areas of traditional investment (such as in conventional assets geared to a serious major foreign competitor), it would be difficult to invest in "leap ahead" capabilities that cover a wider solution space.

A key part of capabilities-based planning is to identify and analyze key risks so as to inform leadership decisions on resource allocation. In the recent legacy process, risk judgments were typically made at lower levels, where there is often a nontransparent inclination to buy down risk (at great expense) in some areas while ignoring other types of risk. Capabilities-based planning aims to provide more meaningful information and to better inform strategic-level decision making.

There are multiple dimensions of risk judgments. Each type of risk consideration reflects an attempt to step back from the defense program and to view the program as a nonadvocate with a top-level perspective would view it. This approach is antithetical to the defend-the-program legacy process sometimes employed by the Services for the purpose of program advocacy. Four main areas of risk particularly pertinent in the context of weapons systems and force structure are as follows:[5]

- Strategic risk
  —In what areas do the Services' programs diverge from the DOD's strategy and the Secretary of Defense's priorities?
  —If developments in the security environment indicate that incorrect assumptions were made about a threat, to what extent are the capabilities of our planned force adaptable to meeting new threats?
  —Can the defense program be substantially changed in time to meet an unexpected need?
- Current versus future risk
  —Is our current and near-term investment crowding out critical science and technology investments needed to address more advanced future threats?
- Program-schedule risk
  —Which individual programs will slip, overrun, or both? What is the potential impact?
- Technical risk
  —Will individual platforms and systems provide the intended or needed capability?

[5]These four areas of risk (with different subsets of questions) also apply to manpower, training, logistics, and other functional areas essential for deployed forces to be able to successfully carry out missions assigned to combatant commanders.

The risk-assessment process requires an extensive feedback loop between staff and leadership. Early analysis should feed leadership-level judgments on risk-tolerance levels, key objectives, and priorities. Leadership risk judgments in turn should be communicated at the start of the process and, as more knowledge is gained, they should direct or redirect the organization's broader analytic activities. This direction should indicate, at the macro level, those areas of need in which the leadership expects the organization to attenuate risk as well as those areas in which the organization can divest resources or hold steady (and accept any perceived increased risk).

Service-level direction should refine, elaborate, and make more meaningful the Secretary of Defense's DOD-level guidance and should not be a statement of Service priorities unrelated to corporate guidance. As discussed above, the organization should be providing continuous feedback to the leadership on whether the initially stated risk-tolerance levels need to be slightly or even radically changed to be achievable.

In summation, the key elements and criteria associated with the capabilities-based approach of confronting uncertainty, expressing risk in meaningful terms, exploring a wide range of possible solutions, and supporting leadership decision making are shown in Figure 2.3.

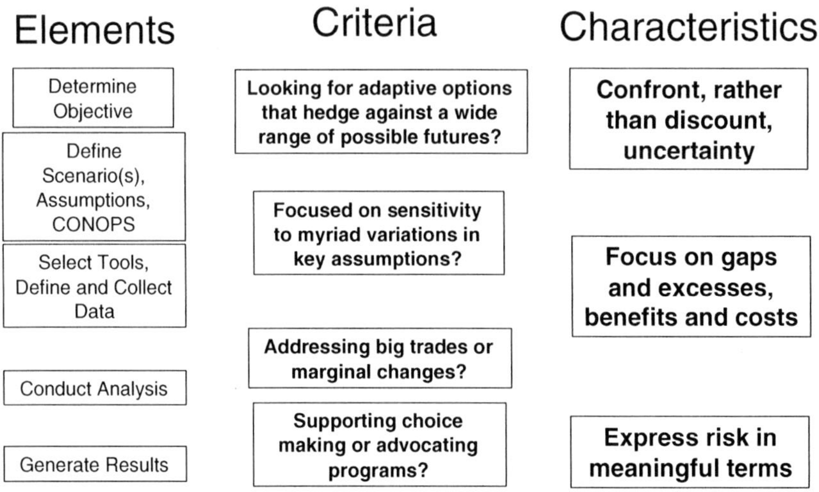

FIGURE 2.3 Key elements, criteria, and characteristics for carrying out capabilities-based planning. NOTE: CONOPS, concept of operations.

3

Review of the Navy's Analytical Processes and Methods

THE COMMITTEE'S APPROACH

The committee framed its review of the Navy's efforts to implement capabilities-based planning (CBP) by asking about the Navy's (1) conceptual framework for CBP, (2) its analytic framework, (3) its explicit attention to future building blocks, and (4) implementation at the level of personnel and organizations.

Conceptual Framework for Capabilities-Based Planning

Basic Ideas

As discussed in Chapter 2, the CBP approach is fundamentally about planning under uncertainty by emphasizing flexibility, robustness, and adaptiveness, while doing so within an economic framework. That is, choices must be made about how much risk of various types is tolerable, how to exploit opportunities for efficiency and effectiveness, and how to live with the budget that is finally decided upon by national authorities. The capabilities in question are to be outputs—measures of the ability actually to execute tasks, missions, and operations. These capabilities should also be conceived as *joint* capabilities, even though in some instances a particular joint capability may effectively be a Service capability (e.g., undersea surveillance).

"Planning under uncertainty" is not a mere phrase being emphasized; it is the essence of CBP, which recognizes that the United States cannot reliably predict how its military forces will be used—against whom, for what purpose, and in what circumstances. Nor can it reliably predict precisely how conflict situations will evolve or how well each and every system and operation will work. Hedging is essential. But hedging is costly. How much is enough? And what framework is used to judge?

Simple Tests

Some simple tests can indicate whether a Service is applying CBP. One is whether the emphasis is on achieving capabilities rather than, as in prior periods, platforms and weapons systems. A second test is whether options for achieving capabilities are joint and whether trade-offs cross Service boundaries where appropriate. A third test is whether risk is considered, in its various dimensions. And, last but not least, are assessments accomplished within an economic framework, which includes identifying funding sources for additions that would otherwise defy the fiscal guidance?

Analytic Framework

Given a conceptual framework, an organization also needs a suitable analytic framework to conduct CBP, preferably one that is widely understood and that provides for the following:

- An understanding of capability needs at the mission or operation level;
- An understanding of *aggregate* capability needs (for theater and multitheater challenges);
- The development and assessment of options for providing needed capabilities, including options that maintain the overall funding level specified by fiscal guidance; and
- The assessment of options and trade-offs in an integrative portfolio-management structure suitable to Chief of Naval Operations (CNO)-level review.

Because issues arise at different levels (e.g., strategic, campaign, and mission), it follows that the analytic framework must be hierarchical, with a clear logic trail from the high-level constructs down to those in which one can see the critical components and subcomponents of capability that make operations successful. The relationships among levels of analysis cannot merely be asserted on the basis of implicit assumptions; instead, they must be derived from thoughtful, explicit analysis with conscious trade-offs.

Consideration of Future Building Blocks

In domains in which one seeks flexible, adaptive, and robust capabilities, effective solutions typically depend on developing appropriate capabilities as building blocks. These exist in the realm of systems and platforms, organization, and operations. Planning for the future in an uncertain era of dynamic change implies rethinking—and probably transforming—the building blocks. For the Navy, this will likely mean new strike groups, different concepts of manning, and new joint operations or ways of conducting old ones.

Implementation—Moving Toward First-Class Analysis

Finally, conducting CBP well will require first-class analysis. Achieving this objective involves institutional issues and has major implications for staffing, organization, and reward systems.

With the background presented in this and the preceding subsections as a framework, the remainder of the chapter addresses the issues in the same order discussed above and provides the committee's assessments and recommendations.

THE CONCEPTUAL FRAMEWORK

Does the Department of the Navy have a sound, top-level conceptual framework to guide capabilities-based planning? To address this question, the committee drew primarily on the following documents and briefings: (1) "Sea Power 21 Series" articles from *U.S. Naval Institute Proceedings*,[1] (2) *Naval Transformation Roadmap 2003*,[2] (3) a set of briefings presented to the committee by the Office of the Deputy Chief of Naval Operations (DCNO) for Warfare Require-

---

[1]ADM Vern Clark, USN, Chief of Naval Operations. 2002. Sea Power 21 Series—Part I, "Projecting Decisive Joint Capabilities," *U.S. Naval Institute Proceedings*, October; VADM Mike Bucchi, USN, and VADM Mike Mullen, USN. 2002. Sea Power 21 Series—Part II, "Sea Shield: Projecting Global Defensive Assurance," *U.S. Naval Institute Proceedings*, November; VADM Cutler Dawson, USN, and VADM John Nathman, USN. 2002. Sea Power 21 Series—Part III, "Sea Strike: Projecting Persistent, Responsive, and Precise Power," *U.S. Naval Institute Proceedings*, December; VADM Charles W. Moore, Jr., USN, and LtGen Edward Hanlon, Jr., USMC. 2003. Sea Power 21 Series—Part IV, "Sea Basing: Operational Independence for a New Century," *U.S. Naval Institute Proceedings*, January; VADM Richard W. Mayo, USN, and VADM John Nathman, USN. 2003. Sea Power 21 Series—Part V, "ForceNet: Turning Information into Power," *U.S. Naval Institute Proceedings*, February; VADM Mike Mullen, USN. 2003. Sea Power 21 Series—Part VI, "Global Concept of Operations," *U.S. Naval Institute Proceedings*, April; VADM Alfred G. Harms, Jr., USN, VADM Gerald L. Hoewig, USN, and VADM John B. Totushek, USN. 2003. Sea Power 21 Series—Part VII, "Sea Warrior: Maximizing Human Capital," *U.S. Naval Institute Proceedings*, June.

[2]ADM Vern Clark, USN, Chief of Naval Operations; and Gen Michael Hagee, USMC, Commandant of the Marine Corps. 2004. *Naval Transformation Roadmap 2003: Assured Access and Power Projection from the Sea*, Department of the Navy, Washington, D.C.

ments and Programs (N70)[3] and the Assessments Division of the Office of the DCNO for Resources, Requirements, and Assessments (N81);[4] the Naval Air Systems Command (NAVAIR);[5] and the Office of the DCNO for Manpower and Personnel (N1).[6] The relationships of these Navy CBP efforts to approaches of the Office of the Secretary of Defense (OSD) and the Office of the Joint Chiefs of Staff (OJCS) are discussed in Chapter 4.

This assessment addresses separately the Navy's broad strategic approach, its system level of analysis, and its analysis at the mission and operational level, as in development of Program Objective Memorandums (POMs) responsive to strategic guidance.

Broad Strategic Approach

The committee concludes that the Department of the Navy has done a creditable job in laying out a broad strategic approach and has gone on to delineate sensibly the special responsibilities that the maritime Services have in national strategy and joint operations. The Navy's approach is organized at the top level in terms of Sea Shield, Sea Strike, Sea Basing, and the enabling "glue" of FORCEnet, as indicated in Figure 3.1.[7] These are supported by what are termed Sea Trial, Sea Warrior, and Sea Enterprise.

Were the planning to stop with this top level, it would produce little more than good viewgraphs, but the Navy has put considerable effort into assuring that all of the important functions of the department are mapped into this structure and that useful decompositions exist down to meaningful levels of detail (see the next subsection). Such breakdowns are always imperfect because of crosscutting factors, but the committee was satisfied that the structure largely makes sense. The structure will probably change over time as the Navy gains experience with the decomposition and makes adjustments, but the approach is sensible. This said, the committee notes that the approach is quite different from that being used in OSD and the Joint Staff (in Chapter 4, see the discussion of the Joint Capabilities Integration and Development System (JCIDS)), which involves functional capa-

---

[3]CAPT Terry McKnight, USN, N70, "Naval Capabilities Development Process," presentation to the committee, July 27, 2004, Woods Hole, Mass.

[4]LCDR Kenneth Masson, USN, N815, "Capabilities Based Planning," presentation to the committee, July 27, 2004, Woods Hole, Mass.

[5]Patrick McLaughlin, NAVAIR, "Naval Analytical Capabilities and Improving Capabilities-Based Planning," presentation to the committee, July 28, 2004, Woods Hole, Mass.

[6]Richard Robbins, N1Z, "N1 and Capabilities-Based Planning," presentation to committee members, July 21, 2004, Navy Annex, Washington, D.C.

[7]CAPT Terry McKnight, USN, N70, "Naval Capabilities Development Process," presentation to the committee, July 27, 2004, Woods Hole, Mass., slide 10.

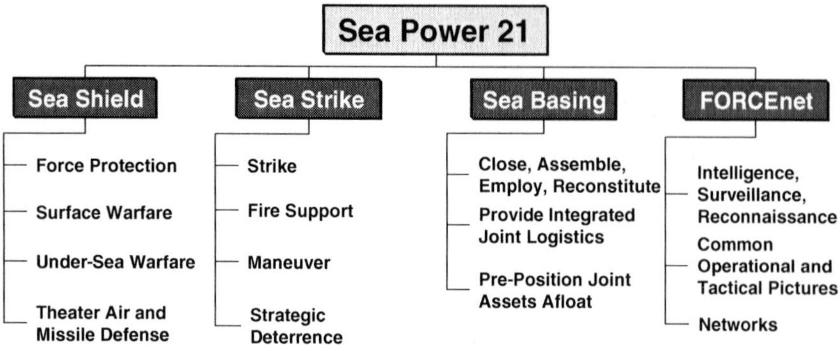

FIGURE 3.1 Top-level components of Sea Power 21. SOURCE: CAPT Terry McKnight, USN, N70, "Naval Capabilities Development Process," presentation to the committee, July 27, 2004, Woods Hole, Mass., slide 10.

bility areas identified as focused logistics, battlespace awareness, force application, force protection, command and control, network-centric operations, training, and force management. It is important that the Navy have clear mappings from its decomposition to the Department of Defense's (DOD's) functional capability areas if it is to participate and compete effectively in overall DOD planning. The mapping issue is nontrivial because the Navy capabilities, natural in an operations-oriented decomposition, depend on a number of the functional capabilities in JCIDS.

The committee is also convinced that at the highest levels of the Navy and the Marine Corps there is a commitment to jointness—not merely to offer lip service to it but because jointness is a fundamental aspect of overall transformation for the new era in which the United States finds itself. At that highest level, as reflected in the core documents, it is appreciated that warfighting will almost always need to be joint in the future. Even under those circumstances, however, enormous responsibilities will continue to devolve upon the maritime commanders.

Finally, the higher-level documents all reflect a commitment to flexibility, adaptiveness, and robustness. This is perhaps not surprising, since the Navy and Marine Corps have traditionally emphasized these qualities to a greater extent than have the Army and Air Force, which became more captive to planning for particular war scenarios.

In contrast, the committee was not persuaded that the translation of higher-level intentions into lower-level processes and practices is going well, as discussed below in the subsection "Operational Analysis for the Department of Defense and Office of the Chief of Naval Operations" and in the next major section, "Analytic Framework." First, however, work at the system-command level is discussed.

System-Level Analysis

The committee was generally impressed by the presentations made by representatives of naval organizations two levels down from the highest level. These presentations had been generated by the Navy's Space and Naval Warfare Systems Command (SPAWAR), the Naval Sea Systems Command (NAVSEA), and the Naval Air Systems Command. Here the committee saw evidence of managerial rethinking about organization, process, and products to support CBP. For example, in the briefings cited, the committee saw reference not only to analytical work on capabilities, but also to life-cycle costs, and business-case assumptions. The Navy has even reorganized to operate what it calls a Virtual Systems Command to increase agility and integration.[8] Illustrative discussion of the Virtual Systems Command's analytical process reflected a systems-engineering perspective, with the inclusion of systems of systems and connections to mission-capability packages and the discrete "things" that end up being line items in budgets. It also suggested determination not only to identify overlapping capabilities but to distinguish between desirable and undesirable redundancy and to identify both capability gaps and trade-offs. At least in a quick-look review, this class of work appeared to be professional and responsive to the new paradigms of CBP. Whether the Virtual Systems Command will work out is, of course, something that only experience will show.

The committee also heard a briefing from the Navy Warfare Development Command (NWDC),[9] which addressed rather extensive fleet-based experimentation to support near-term assessments closely related to filling recognized capability gaps (e.g., those against small-boat attacks). This effort reflected a laudable Navy decision to reemphasize fleet-level experimentation and the accumulation of substantial empirical and analytical data. The experiments described, however, were all focused on the near term. Although all of them were clearly desirable and important, the committee was concerned that the effort might remain too exclusively concerned with near-term, incremental issues. The Navy leadership will wish to review issues of balance over time.

---

[8]Patrick McLaughlin, NAVAIR, "Naval Analytical Capabilities and Improving Capabilities-Based Planning," presentation to the committee, July 28, 2004, Woods Hole, Mass.

[9]Wayne Perras, Technical Director, Navy Warfare Development Command, "What We Do/Who We're Doing It For," presentation to the committee, July 27, 2004, Woods Hole, Mass.

Operational Analysis for the Department of Defense and Office of the Chief of Naval Operations

In contrast to the experience described above, the committee found many reasons for concern at the level between top-level guidance and systems command (SYSCOM)-level work. Here the committee observed severe disconnects between top-level intentions and reality in the ranks. When the committee asked about excursions and exploratory analysis around baseline assumptions, briefers reported that there had been very little done. Thus, while some viewgraphs had been changed to be consistent with CBP, much of the ongoing work still had the problems of the previous era, particularly those surrounding point-scenario analysis. Although the problems seen by the committee may have been temporary, they appeared more likely to be chronic. If so, the Navy should recognize these as being systematic, indicating deep-seated issues, and act accordingly. Corrective measures will take much more than top-level documents, because staff take their lead from a myriad of actions and priorities expressed in the course of time. These problems are discussed more fully in the next section.

THE ANALYTIC FRAMEWORK

Understanding Needs at Mission and Operation Levels

Description of Analytic Approach at Mission and Operation Levels

The Department of the Navy's core documents include useful decompositions from high-level components (e.g., Sea Strike) down to meaningful levels of detail. It is always a matter of judgment how far to carry such breakdowns. As one goes into more detail, issues and tasks become increasingly well defined and challenges become more explicit. However, excessive decomposition also generates a morass of detail that is not useful for higher-level planning.[10] And, to make things worse, it can introduce biases by "hard-wiring" the way in which missions and higher-level tasks are to be performed. In capabilities-based planning, it is desirable to stop decomposing before that happens or, at least, to carry along alternative decompositions reflecting alternative concepts of operation.

As can be seen from Figure 3.2, the Department of the Navy's objectives-to-challenge decomposition structure goes down three to four levels. For example, in the FORCEnet component (close to bottom right), it goes down to the level of "Detect and Identify Targets/Moving Land Targets." This level of specificity is

---

[10]Detailed decomposition is, however, valuable for defining the myriad detailed tasks that must be mastered, supported, and coordinated. All Services are required to prepare detailed decomposition, and the results are published as the Unified Joint Task List by the Joint Staff.

useful for highlighting an important mission that is very different from other detect-and-identify missions, and one that the Navy has not traditionally emphasized. If the mission to engage moving land targets had been left implicit, it might not receive adequate attention.

By and large, the committee concludes that the decomposition shown in Figure 3.2 is suitable as a top-down depiction. In particular, it has enough detail so that responsibilities for follow-up work can be assigned meaningfully. And, although there are scores of capability areas indicated (counting at the lowest level), the number is small enough to be managed. What matters, of course, is that for each one of these capability areas, the Department of the Navy does in-depth analysis to assess needs, capabilities, and improvement options. The committee could hardly review or assess that effort in a cursory review. However, Figure 3.3 illustrates how, for a large subset of items in Figure 3.2, the Navy has sought to assess capabilities versus time. For example, in the highest bar ("Neutralize Submarine Threats"), the Navy's assessment is that the ability to neutralize submarine threats will improve from poor (black) to marginal (gray) within the time period shown (roughly through 2020). In contrast, much better progress is projected for countering minefields (by what mechanism was not made clear to the committee). The assessments were the result of subjective warfighter estimates, informed also by results of POM-06 campaign analyses and mission-level analyses. The process used to obtain the estimates was neither rigorous nor satisfactory to participants, but it was a systematic first effort that can be refined with time.

The analytical approach being employed, then, appears to be that of using the decompositions, examining needs and capabilities in each area, and projecting changes over time in high-level depictions. That approach is reasonable, and was also consistent with the need in CBP to go to mission level rather than merely reporting results of theater-level campaigns in particular scenarios. The committee was surprised by the results shown in Figure 3.3 (almost none of the assessments improve beyond marginal (gray) or poor (black), thereby suggesting that a hard look at the criteria used would be appropriate), but the assessment process was at least a good beginning for something that can be much enriched over time.

So far, so good. Unfortunately, the committee's assessment was that many problems exist at the next level of analytical detail, as discussed below.

Assessment of Analytic Framework for Mission and Operation Levels

In assessing the Navy's mostly implicit analytic framework, the committee drew on its experience and looked for generic problems that often beset analysis that is intended to, but actually does not, support capabilities-based planning. OPNAV will wish to review the situation when this report emerges, but *the generic problems* are as follows:

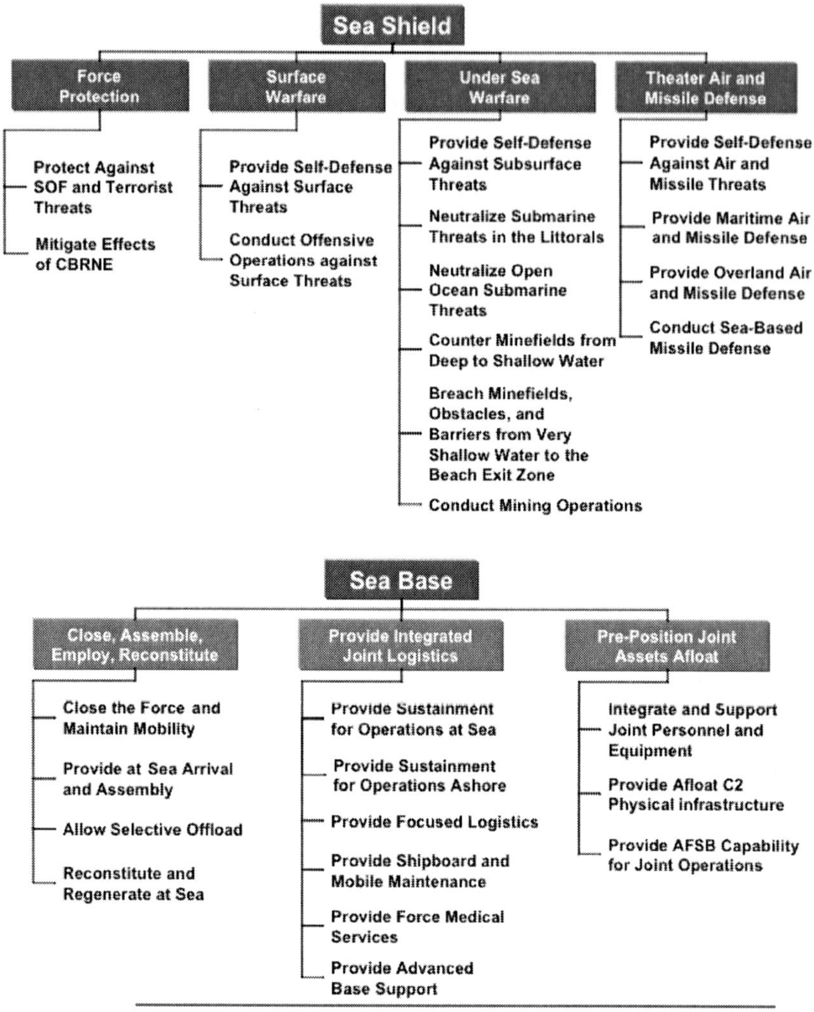

FIGURE 3.2 Decomposition of capability needs. SOURCE: CAPT Terry McKnight, USN, N70, "Naval Capabilities Development Process," presentation to the committee, July 27, 2004, Woods Hole, Mass., slides 11-14. NOTE: SOF, Special Operations Force; CBRNE, chemical, biological, radiological, nuclear, explosives; C2, command and control; AFSB, afloat forward staging base; PNT, precision, navigation, and timing.

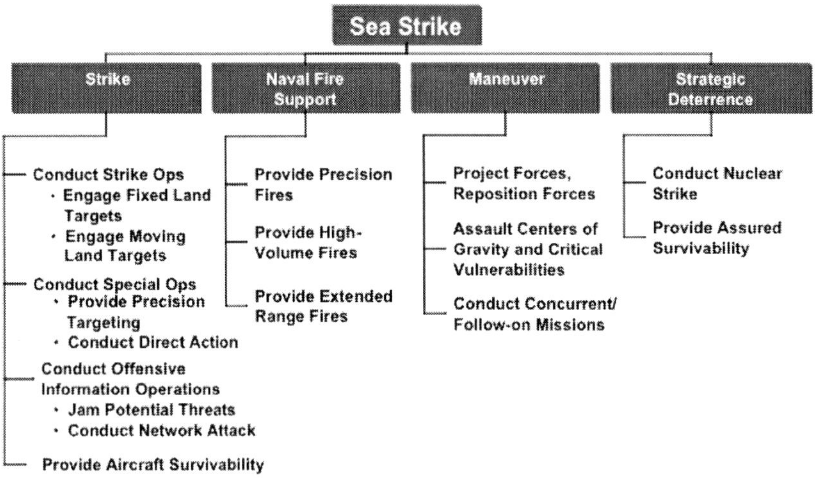
Sea Strike
Strike
Naval Fire Support
Maneuver
Strategic Deterrence
Conduct Strike Ops
• Engage Fixed Land Targets
• Engage Moving Land Targets
Conduct Special Ops
• Provide Precision Targeting
• Conduct Direct Action
Conduct Offensive Information Operations
• Jam Potential Threats
• Conduct Network Attack
Provide Aircraft Survivability
Provide Precision Fires
Provide High-Volume Fires
Provide Extended Range Fires
Project Forces, Reposition Forces
Assault Centers of Gravity and Critical Vulnerabilities
Conduct Concurrent/ Follow-on Missions
Conduct Nuclear Strike
Provide Assured Survivability

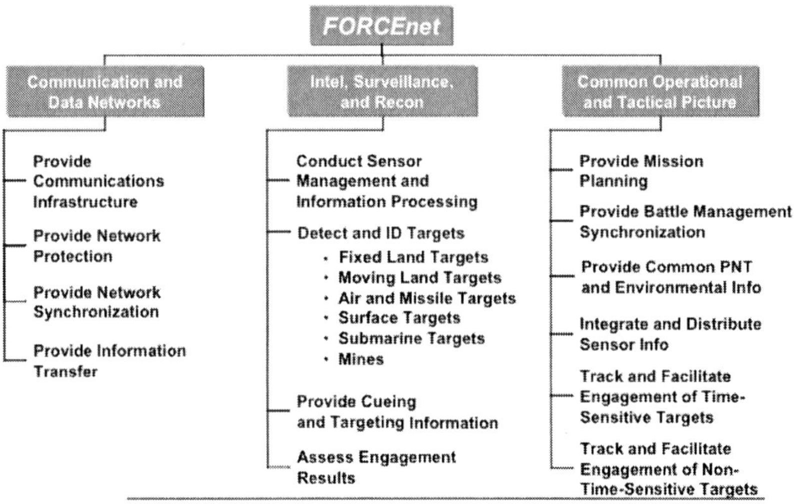
FORCEnet
Communication and Data Networks
Intel, Surveillance, and Recon
Common Operational and Tactical Picture
Provide Communications Infrastructure
Provide Network Protection
Provide Network Synchronization
Provide Information Transfer
Conduct Sensor Management and Information Processing
Detect and ID Targets
• Fixed Land Targets
• Moving Land Targets
• Air and Missile Targets
• Surface Targets
• Submarine Targets
• Mines
Provide Cueing and Targeting Information
Assess Engagement Results
Provide Mission Planning
Provide Battle Management Synchronization
Provide Common PNT and Environmental Info
Integrate and Distribute Sensor Info
Track and Facilitate Engagement of Time-Sensitive Targets
Track and Facilitate Engagement of Non-Time-Sensitive Targets

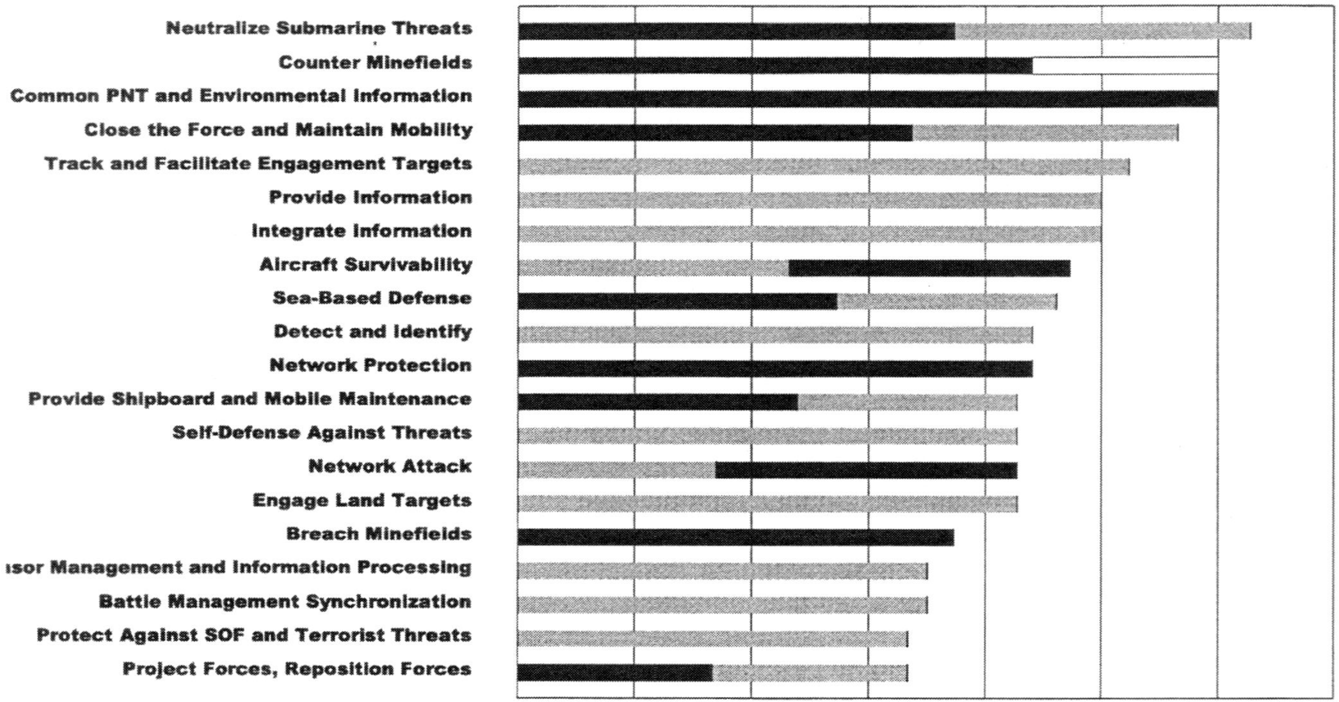

FIGURE 3.3 Projected capability versus time (roughly through 2020) for each of the capability areas. SOURCE: Adapted from CAPT Terry McKnight, USN, N70, "Naval Capabilities Development Process," presentation to the committee, July 27, 2004, Woods Hole, Mass., slide 25. NOTE: PNT, precision, navigation, and timing; SOF, Special Operations Force. Key: Poor – black; Marginal – Gray; Good – White

- *Trivializing uncertainty issues* by examining a variety of name-level (or specific) scenarios (e.g., in a DOD context, the "WestPac" scenario), but approaching each such scenario in the traditional, myopic way, holding constant all of the many assumptions that distinguish one case from another within the scenario. For the purposes of CBP, there are often more important variations across cases within a given name-level scenario than across name-level scenarios.
- *Trivializing exploratory analysis*, which is a core element of CBP, by conducting only a few excursions with one or a few assumptions changed (typically in organizationally comfortable ways), while other assumptions are held constant as though certain.[11]
- *Focusing on scenarios preferred by an organization* through emphasis on detailed scenario assumptions that stress the particular organization's issues and dramatize its role, without presenting a more holistic representation of how the organization fits into a larger enterprise and its activities.
- *Relying upon large, complex, and inflexible models and databases*, which often bury issues and preclude exploration.

The Department of the Navy analysis (of a scenario of a major war of a traditional sort) briefed to the committee at its July 2004 workshop reflected serious problems in each of the categories listed above. There was very little discussion or appreciation of uncertainty, except sometimes in boilerplate slides. Worse, this vacuum seemed to be regarded by the analytic staff as the norm. Indeed, it was claimed that there was insufficient time to do many excursions. The committee finds this trivializing of uncertainty issues quite troubling and inconsistent with fundamental tenets of capabilities-based planning. From the committee's own collective experience this is a general problem, not one confined only to the Navy staff. Analytical organizations in the DOD frequently do not have sufficient breadth in the tools, staff, and experience needed to carry out such broad analysis adequately, particularly in the current threat environment. This lack needs to be remedied quickly.

To make matters worse, as discussed in Chapter 4, the scenarios used in the OPNAV analysis did not include the DOD standard scenarios established by the OSD and the OJCS. Instead, they were scenarios and case assumptions within those scenarios chosen to stress naval capabilities in particular. It is entirely appropriate that such cases be examined, but not by themselves. It was difficult to avoid the conclusion that a reason for choosing the scenarios and cases was the

---

[11]An older example of this problem was the tendency of DOD studies to assume that adversaries were highly capable and motivated, thereby biasing analysis against concepts of operations calling for rapid action by small forces. A more recent example was the analysis for the Iraq war that uncritically accepted assumptions about the nature of the postwar environment rather than laying out starkly the consequences of different assumptions. In the Navy context, key assumptions often relate to the availability of other-Service assets, concepts of operation, and enemy strategy.

classic motivation of "making the Navy case" in the battle for funds rather than objectively describing the analytic landscape.

Again, it is entirely appropriate that the Navy identify the cases that cause it stress and make their implications known to policy makers. However, there is an obligation to work and report on DOD standard cases, and to put Navy-stressful cases into analytical perspective. Moreover, first-rate capabilities-based planning in the modern era must be conceived in a joint context, with serious steps taken to identify how to exploit jointness effectively in mitigating problems that would classically have been seen as Service-level problems. If, for example, some Air Force bases that could potentially be used are assumed unavailable, it does not follow that a war must now be fought from carrier strike forces alone. Other options need to be addressed: What could be accomplished with long-range bombers (B-1s, B-2s)? What regional bases might reasonably be available and feasible to protect during operations? How would that change as a function of political-military considerations and the availability of ballistic-missile defenses?

In short, there was a stark disconnect between the CNO's stated intentions with respect to jointness and practice in carrying out these intentions. On the one hand, top-level documents and briefings by the Assessments Division of the Office of the DCNO for Resources, Requirements, and Assessments (N81), among others, emphasized jointness and the need to consider a broad range of both scenarios and cases within them—a major departure from analysis in past years. On the other hand, OPNAV analytic work looked more like business as usual: narrow and isolated from joint analytical efforts. The causes of this disconnect are not clear to the committee. Although the committee is convinced that the Navy leadership is fully committed to jointness, it nonetheless believes that Navy leadership needs to do more to carry out its intentions.

Understanding Aggregate Capability Needs

Description of and Rules of Thumb for Aggregation

Even if Department of the Navy assessments of mission- and operation-level capability needs were perfect, the problem would remain of how to aggregate these needs. *How much* antisubmarine warfare (ASW) capability is needed? *In how many* places would ASW be conducted simultaneously against troublesome threats (e.g., diesel submarines)? *How much* ground-strike capability is needed in an era of precision weapons when such weapons can be called to bear from multiple sources (land-, sea-, and air-based), the sources can communicate rapidly, and all are locked together in the same time-and-space coordinate system?

In addition to questions about how much capability is needed, another aspect of understanding aggregate needs is characterizing how quickly one needs them and how quickly missions need to be accomplished. By way of analogy, the need to be able to interdict ground forces has been recognized for many decades. It was only when the need arose to halt enemy ground forces *quickly* (within days rather than weeks, and within only kilometers rather than hundreds of kilometers of lost territory) that the need developed intensity and posed serious challenges. A number of studies were done between 1993 and 2001 to better understand what could be accomplished in this regard and how it translated into requirements for forward-deployed forces, deployment rates, per-sortie effectiveness, minimization of the time required to suppress air defenses, and so on. There was no single answer, because *how much is enough* depended on assumptions about usable warning, regional access to bases, the size of the enemy force, its rate of advance during a continuing interdiction campaign, and so on. Nonetheless, some conclusions were evident, such as the necessity of having forces available on D-Day and of being able to use them immediately, without a lengthy defense-suppression campaign of the type that had been assumed in most studies in the early-to-mid-1990s.

Some of the significant issues today, with obvious implications for assured access and Navy requirements, include whether to assume the availability and continued viability of regional bases for the Air Force to use, the warning time available to the Air Force, and possible political constraints on the use of those bases in times of crisis (particularly in Japan).

These are classic problems. They are also the problems for which strategic planning guidance is intended to be both helpful and, yes, directive. The Navy needs to provide objective analysis that describes sharply the potential consequences of overly optimistic assumptions about the early use of massive Air Force assets, but that analysis also needs to describe clearly (1) how maritime operations could be enhanced by the appropriate use of other-Service assets (long-range bombers, surveillance systems, conventional ground forces, and special forces) in those stressful cases; (2) how maritime forces would be supportive rather than dominant in other important cases (e.g., when they might be critical in the first days for theater-opening purposes, but less so thereafter); (3) the decreasing plausibility of the maritime Service having to deal with particularly stressful circumstances simultaneously in multiple theaters; and (4) how maritime Services would cope in plausible simultaneous operations in other theaters if focused upon operations in the principal theater.

Some rules of thumb for aggregation (roll-up) are as follows:

- When working at the mission level (see the preceding subsection), mission-system analysis should be required—by which is meant that *all* of the critical components of capability must be healthy or the overall assessment of that capability should be ranked poorly.[12]

---

[12]This is akin to using a multiplication scheme rather than a linear weighted sum in aggregating upward.

• When addressing force structure, bad cases such as simultaneous conflicts *should* be taken seriously, but "requirements" should be developed in recognition that crises must often be addressed by the nation sequentially for many reasons, and that in the event of simultaneous crises, delay tactics may be appropriate.[13]

• Similarly, while it is important and necessary to emphasize rapidity of response, this can also be overdone. The value of timeliness is highest for the first increments of capability; it diminishes thereafter—especially when it is recognized that *real* conflicts almost invariably unfold much more slowly than in studies, for numerous reasons having to do with decision making, ambiguity, and the need to avoid the high operational risks associated with committing to a course of action before it can be sustained (as in the march to the Chinese border in 1950). The point here is not to denigrate rapid response, the importance of which can be immense, but rather to avoid excess. Although it might appear that the Navy would get more return for its investment if it stressed rapid deployment of eight carrier strike groups, questions such as that involving the trade-off between slipping the arrival time of the last two carrier strike groups, say, and the ability to retain personnel, the ability to invest in experiments with new kinds of strike groups, and so on, need to be addressed.[14]

Assessment of Analysis of Aggregate Capability Needs

As is being widely discussed across the entire Department of Defense, there are serious problems in assessing overall force-structure needs.[15] Some of these are inherent consequences of uncertainty, while others reflect various types of parochialism. The committee is encouraged that parts of the Department of the Navy are now vigorously pursuing concepts for improving effectiveness that were previously off the table. These include the Fleet Response Plan (with higher readiness levels that enable more flexible response), rotational crews, and the creation of expeditionary strike groups (ESGs).

These inventive solutions seem very much a response to the demands of the

---

[13]In DOD-level planning, this has sometimes been discussed under the rubric of "strategies" of "win-win" versus "win-hold." The current strategy goal described as 1-4-2-1 is the result of considerable debate on such matters.

[14]Some of the analysis essential in answering these questions is being done. The committee heard Ariane L. Whittemore, N4, "Fleet Response Plan and the Integrated Readiness Capability Assessment (IRCA)," presentation to the committee, July 28, 2004, Woods Hole, Mass.

[15]Such discussions are occurring, for example, in meetings devoted to the implementation of capabilities-based planning within the DOD. See also, "Initiation of a Joint Capabilities Development Process," memorandum from the Secretary of Defense, October 31, 2003, to the Service Secretaries, Chairman of the Joint Chiefs of Staff, and others, setting a goal and issuing guidance to "achieve a streamlined and collaborative, yet competitive, process that produces fully integrated joint warfighting capabilities."

DOD's strategic objectives for defending the homeland and conducting military operations. As so often happens in Service-based analysis, however, the innovations did not always have visibility in the capabilities analysis that was probably seen as relating to Navy "requirements" and in the battle of the budget. The committee would like to have heard about a capabilities-based exploration of the possible trade-offs between carrier strike group (CSG) force structure and a heightened readiness posture, or even a more radical approach such as trading off CSGs against ESGs and aerospace expeditionary forces.

Metrics

Establishing metrics is another major challenge throughout the DOD, as emphasized by the Secretary of Defense.[16] Managers want metrics, as do oversight groups such as the U.S. Congress, the Government Accountability Office, and the Office of Management and Budget. Unfortunately, using the wrong metrics can be seriously counterproductive, and metrics that can be used internally in sensible ways can be used mischievously by organizations.

Following are some general principles for metrics in aggregate-level planning, such as force-structure studies. The first principle is to develop metrics as spin-offs of operational analysis, so that the metrics fit naturally into analysis of capabilities to accomplish missions and operations. This is in contrast with emphasizing simpler, bean-counting types of metrics such as numbers of platforms.

Goals for metrics should be based on analysis and realistic assumptions about technical feasibility. They should not be established ad hoc by decision makers who would like to be able to legislate magic. If metric goals are unrealistic, they corrupt not only subsequent analysis (who wants to go forward with an analysis showing that the boss's goal cannot be reached!), but also broader aspects of the organization's management.[17]

The metrics, however, should build in the emphasis of capabilities-based planning on flexibility, adaptiveness, and robustness. This can be done, for example, by reporting mission outcomes in an exploratory analysis across cases rather than reporting the mission outcome for some allegedly representative point case.

An example for a naval application might be that of assessing the distance required to stop an enemy's maneuver force with a combination of aircraft and

---

[16]See Charles Kelley, Paul Davis, Bruce Bennett, Elwyn Harris, Richard Hundley, Eric Larson, Richard Mesic, and Michael Miller, 2003, *Metrics for the Quadrennial Defense Review's Operational Goals*, RAND, Santa Monica, Calif.

[17]The committee's concern is motivated in part by other Services having postulated unachievable capabilities, which then led to analysis "supporting" (but really just saluting) the postulates, wheel spinning, and loss of time from pursuing poor courses of action. Goals should be ambitious, but should also be rooted in the real world.

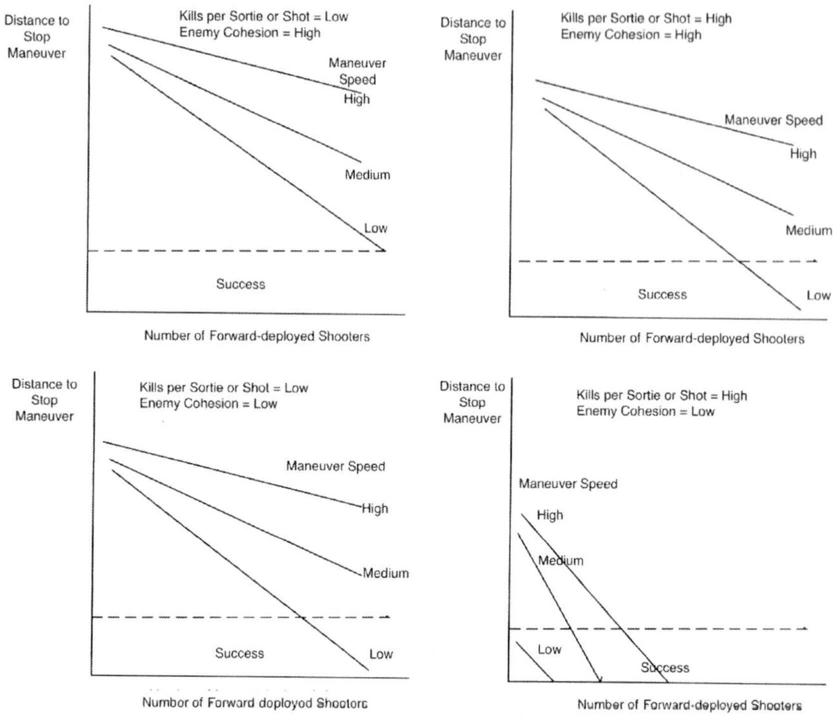

FIGURE 3.4 A notional example of seeing a metric as outcome across a case space (exploratory analysis). See the discussion of outcomes in text.

missiles. Instead of looking at a particular point scenario, CBP would treat at least the following as key variables: (1) the number of forward-deployed aircraft and missiles (shooters), (2) the speed of the enemy's maneuver, (3) the effectiveness per missile shot or aircraft sortie, and (4) the enemy's cohesion.[18]

Figure 3.4 illustrates schematically how analysts could look at the simultaneous variation of these four variables. In the top left quadrant of the figure, the enemy moves a long distance before being stopped, because the maneuver speed is high, kills per sortie or shot are low, and the enemy's cohesion is high. Even having a large number of shooters available would not result in success. Success

[18]Actual curves would not be linear. For exploratory analysis of the counter-maneuver problem (albeit, one that preceded the war in Iraq), see Paul K. Davis, Jimmie McEver, and Barry Wilson, 2002, *Measuring Interdiction Capabilities in the Presence of Anti-Access Measures: Exploratory Analysis to Support Adaptive Planning for the Persian Gulf*, RAND, Santa Monica, Calif.

(stopping the attacker after only a short distance, as indicated by the horizontal dashed line) could be achieved in several ways, as shown by results in other quadrants of the figure. If kills per sortie or shot were increased (top right), then success would be achieved, but ony if maneuver speed were low. Similarly, if the enemy's cohesion were low, so that maneuver stopped with relatively light attrition, success could be achieved (bottom left). But if effectiveness were high and enemy cohesion were low, success could be achieved even if the number of forward-deployed shooters were smaller (bottom right).

As of July 2004, it appeared to the committee that Navy studies under the rubric of capabilities-based planning did not typically have sound and sophisticated metrics, particularly metrics reflecting uncertainty. This lack appeared to be less true of more detailed analysis, such as was briefed to the committee by the system commands, but no strong conclusions can be drawn.

Characterizing Capabilities

One aspect of capabilities-based planning is keeping track of the multiple dimensions of capability and how programs project improvements in it. Figure 3.5 illustrates this type of effort with a purely notional, spider-chart depiction. It characterizes the Navy's capabilities along several axes: "Control seas," "Assure early access," "Maintain presence," "Project force inland," "Defend homeland from missile attack," and "Defend allies and deployed forces from missile at-

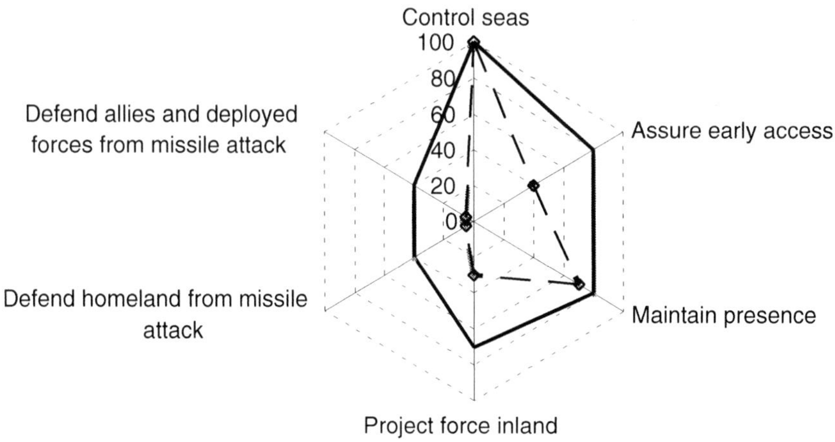

FIGURE 3.5 Notional capabilities-based planning depiction of present Navy capabilities (heavy dashed line) and future goals (heavy solid line). NOTE: Numbers represent percentages.

tack." The inner, dashed, contour suggests that capabilities are strong for two axes—"Control seas" and "Maintain presence"; fairly strong for another—"Assure early access"; and not very strong for the others. One possible goal for future Navy capabilities would be to achieve the outer, solid contour. That would require additional emphasis on ballistic-missile defense, early access, and the ability to project force inland even in difficult circumstances.

General Attributes of Rigorous Analysis

Another major concern of the committee relates to the need for first-rate analysis to be rigorous, documented, transparent, and as objective as possible. Rigor is a matter of degree. High-level decisions on programs and budgets depend on analysis being approximately right and appropriately insightful, not on high precision. Nonetheless, viewgraphs do not constitute analysis, nor do viewgraphs plus assertions about how hard people have worked. Good analysts universally acknowledge that the discipline involved in writing down assumptions and working carefully through the logic—that is, generating documentation—is exceedingly important.

The committee's impression is that OPNAV-level analysis, by contrast with rigorous analysis, is more ad hoc, undocumented, and rather opaque on key assumptions, and that it tends to have an advocacy bias and is constructed to focus only on particular issues (e.g., what might be needed to deal with certain bad-case naval scenarios). The committee did not see broad areas of choice with a hard-edged assessment of strengths and weaknesses being presented to the leadership. The OPNAV-level analysis is worthwhile in some respects, but it is not yet of the quality appropriate for senior decision making.

Choosing Among Options in a Portfolio Framework Suitable to Top-Level Needs

Developing the appropriate portfolio views is a complex undertaking that is highly dependent on the particular organization and decision context.

Strategic Planning Versus Operations Research

As suggested above, the current approach of OPNAV to analysis appears to be one of presenting charts that indicate adequacies and shortfalls (see, e.g., Figure 3.3), by capability area, and presenting occasional operations-analysis charts illustrating particular points. For example, the committee was briefed on some interesting work by the Assessments Division of the Office of the DCNO for Resources, Requirements, and Assessments (N81) that examined "Scud hunting problems" in more technical depth and with more operationally realistic assumptions than usual. As a result, different conclusions were suggested about the potential mix of satellites and unmanned aerial vehicles and about the speed needed for air-to-surface missiles.

Such work is useful and has some of the features of portfolio-management-style work in that one can look critically at the various capability areas and see where the greatest shortcomings exist (subject to the appropriateness of the underlying assumptions). That process, in turn, can lead to suggestions about resource allocation. It is, nonetheless, an operations-analysis perspective rather than one ideally suited to resource allocation.

Assessment of the Navy's Portfolio-Management-Style Analysis

Strategic-Level Portfolio Analysis. In the committee's view, portfolio-management-style presentations for the CNO and his top leaders should often have a more strategic, top-down character, and should more explicitly address economic issues. The committee did not see much economics-sensitive analysis, although some had been done in the preparation of the briefings that the committee was given. It seems, however, that the emphasis in the Navy's CBP is exclusively on identifying shortfalls and finding ways to fund them. Notably absent, except at the systems command level, is the search for opportunities to accomplish missions effectively but at less cost, thereby freeing up funds for other purposes.

Zooming. Portfolio work should allow for zooming in on an area as desired, so that the basis for high-level charts can be examined in depth. It appears, however, that there is minimal rigor in the Navy's current assessments and no systematic way to trace the assumptions and logic from a top-level portfolio view to deeper capabilities analysis in which assumptions and their consequences could be seen parametrically. Arguably, this type of ability requires a family-of-models approach.

The Navy *is* working to establish such a family of models, for which the CNO has provided funds. The architecture for what is needed, however, was not clear to the committee. It must include high-quality policy and system analysis, not just more investment in big models and simulations. It should also be tied to real-world data, not just to simulation.[19]

Highlighting Risks. Highlighting types of risk is a key part of portfolio-style analysis. Examples of different types of risk include the following: technical risk (Will a system in development work?), program risk (Will the program slip in time or have cost overruns?), future technology risk (Is the base being laid, in

[19]National Research Council. 2004. *The Role of Experimentation in Building Future Naval Forces*, The National Academies Press, Washington, D.C.

research and development (R&D), for necessary future systems?), and strategic risk (Will the capability developments prove seriously inadequate because of changes in the strategic environment or national policy?).[20]

Adaptive Options. Strategic options should be explicitly adaptive and should hedge against their key assumptions' proving to be wrong. One way for the Navy to do that is to consider seriously the broad range of scenario classes identified in the Strategic Planning Guidance (SPG).[21] Another is to recognize that the SPG itself is a baseline, not a definitive roadmap into the future. Indeed, key assumptions of the SPG will likely change with administrations and with strategic developments in the world. Thus, the Navy's planning should also consider how robust its program would be in the event of such changes. The Navy should have contingency plans for such *possibilities* as (1) a greatly increased emphasis on defending the homeland from missile attacks (e.g., from containerized missiles) and (2) much-greater-than-expected threats to aircraft carriers, even at rather long ranges. These possibilities are offered purely as examples.

It appears that current Department of the Navy work does not include true strategic options (e.g., adaptive options that hedge against events unfolding in unexpected ways). It is too formulaic and too slavishly responsive to CNO and DOD guidance, without providing feedback that might help reaffirm or adjust that guidance.

Implications for Personnel. Presenting broad, discerning, strategic-level analysis for the CNO requires a higher level of analysis than that characteristic of systems analysis or operations research. This broad analysis is in the realm of strategic planning and policy analysis. Current personnel requirements for OPNAV analysis are predominantly limited to capabilities and experience possessed by operations-research-oriented personnel (and even those requirements are often not met). Revised personnel requirements and personnel policy changes would make it more plausible that those chosen would be rewarded with promotion.

---

[20]Portfolio methods are discussed in the following work and references therein: Paul K. Davis, 2002, *Analytic Architecture for Capabilities-Based Planning, Mission-System Analysis, and Transformation*, National Defense Research Institute, RAND, Santa Monica, Calif. Current applications to the Missile Defense Agency highlight these particular risks. Unpublished discussion of the subject arose in a summer 2004 workshop at the Naval War College sponsored by OSD's Director of Net Assessment. Many portfolio-related discussions, of course, can be found in the business literature, some of which are relevant even though the DOD and military Services do not have simple "bottom lines" against which to measure everything. See, for example, Robert S. Kaplan and David P. Norton, 1996, *The Balanced Scorecard: Translating Strategy into Action*, Harvard Business School, Cambridge, Mass.

[21]Department of Defense. 2004. *Strategic Planning Guidance*, Secretary of Defense Donald Rumsfeld, Washington, D.C. (draft). (Classified)

The Need for Assimilating Capabilities-Based Planning Principles in Navy Analytic Processes

The committee based the following recommendation on the assessments in this section. The recommendation will remain applicable until Navy leaders, after reviewing the situation, discover that they have eliminated the disconnects referred to above between leadership intentions and day-to-day analysis.

Recommendation 1: The Chief of Naval Operations should reiterate principles of capabilities-based planning and ensure that they are truly assimilated in Navy analytic processes.

The criteria for implementing Recommendation 1 include the following: The work accomplished should be joint and output-oriented, with the ability to actually execute operations as output. Successful CBP will require analysis over a broader scenario space, extensive exploratory analysis within specified scenarios, development of options both to solve capability problems and to achieve efficiencies, and portfolio-style assessments of those options at different levels of detail. The portfolio-style assessments should assist in making difficult trade-off decisions and should also address various types of risk that Navy leadership must take into account. Strategic options should be adaptive, because world developments and technological developments will undoubtedly force changes—the potential need for which are not much discussed in the DOD's Strategic Planning Guidance.

Although the committee did not discuss tools to support the types of analysis referred to here, it is quite aware that analytic organizations have trouble responding to the demands of good capabilities-based planning. The difficulties are rooted in excessive dependence on large, complex models and related databases; in management demands for detail; and in the ways in which analyses have been framed and conducted. Breaking these molds will not be easy. It will require a family-of-models approach that includes links to war-gaming and experimentation, but that must also include an often-ignored component: "smart," low-resolution modeling and analysis that can support exploratory analysis (grounded in higher-resolution work or empirical data when appropriate) and put a premium on higher-level insights rather than focusing on minutia.

Recommendation 2: The Chief of Naval Operations and the Secretary of the Navy should ensure that the Navy invests in defining and developing the new generation of analytic tools that will be needed for capabilities-based planning.

Some of the attributes needed in tools include the following: agility in low-resolution modeling coupled with the ability to go into greater depth where needed (achievable with a sophisticated family of models and games); the ability to represent network-centric operations well (including publish-subscribe architectures, rather than node-to-node representations); and the ability to deal with challenges such as those that the OSD refers to as disruptive, catastrophic, and nontraditional scenarios.

The committee is aware that the CNO has funded new work on a family of models. It is quite possible, however, that the funds will quickly be exhausted in improvements to "big models" and databases, with little benefit for higher-level capabilities-based planning, as described above. The committee encourages a balanced use of funds, including the potential purchase or use of available off-the-shelf tools.

It is not possible for the committee to make more detailed suggestions here without a more extensive study. The committee notes, however, that examples of the kinds of tools mentioned above have been developed and applied.

FUTURE BUILDING BLOCKS

Flexible, robust, and adaptive capabilities invariably stem from having "building blocks" that can be quickly assembled, tailored, and used in diverse ways. Such capabilities require suitable building blocks, an appropriate command-and-control system, doctrine, and "practice, practice, practice."

The Naval Services have always been relatively good at such things. Carrier battle groups were building blocks tailored to their theater; today's strike groups have evolved a great deal since the Cold War, and the Navy is actively considering an even wider range of employment options. Marines have always done building-block planning, explaining the absence of a standard Marine Expeditionary Force.

Building blocks come in different forms: equipment (e.g., platforms), organization (e.g., carrier strike groups), and operations (e.g., for conducting long-range air strikes or mounting a surprise assault by Marines). Building blocks are also hierarchical. And, in today's world, networking allows more and quicker tailoring and adaptation, as well as leveraging of the capabilities of individual platforms, units, and suboperations.

Overall, the Department of the Navy appears to be addressing the building-block issues vigorously. Problems are likely to occur, however, such as that of allowing important future building-block innovations to slip away when funding becomes tight. For example, funding the full contingent of carrier strike groups and raising their readiness for rapid deployment (up to eight strike groups within a specified number of days) might come at the expense of more actively pursuing non-carrier strike groups or next-generation carriers that would be more difficult for a future adversary to attack. To help reduce the likelihood of such problems, the Department of the Navy should conduct a review of future building-block options and focus on those designed to increase the range of decision options available to the top leadership. This could be accomplished as part of the actions suggested by Recommendation 1.

IMPLEMENTATION—MOVING TOWARD FIRST-CLASS ANALYSIS

The Office of the Chief of Naval Operations has only recently been reorganized, and much more extensive research would be needed to make any useful recommendations about further changes. Thus, the committee did not discuss organizational issues in great depth, instead commenting on problems that could be resolved with incremental changes. The assessment below touches on organizational problems, developing a first-rate analytical staff, the culture needed for such a staff, and links to the DOD and the other Services. Finally, it touches on the temporal issue of what can be done in the short term rather the long term.

Organization

The committee's conclusions about organizational problems are as follows. The logic for the responsibilities assigned to the Deputy Chief of Naval Operations (DCNO) for Warfare Requirements and Programs (N6/N7) on the one hand, and to the DCNO for Resources, Requirements, and Assessments (N8) on the other, is not entirely clear or persuasive, either to the committee or to some of the officers who briefed the committee and talked with its members.

To elaborate, the idea and use of competitive analysis and creative tension are fine, and the intention of generating alternative perspectives is excellent. However, the current competition between parts of N6/N7 and N8 does not appear to be helpful and involves high opportunity costs. Rather than having two alternative, ad hoc versions of any given issue and sets of undocumented analysis, it would be better to have a first-rate job done of objective analysis informed by alternative points of view. The analysis would be more nearly comprehensive, systematic, parametric, questioning of assumptions (even sacred cows), and transparent than today's. Alternative perspectives could be compared by juxtaposing their implications in analysis charts. Consistent with this suggestion, the committee recommends more emphasis on solid, first-rate analyses by a single organization within OPNAV. These could spin off quick-response, ad hoc analyses as needed.

If the Chief of Naval Operations and the Secretary of the Navy (SECNAV) are to have a highly competent analytical organization, it is essential that the organization (1) report directly, or relatively directly, to the CNO/SECNAV, rather than being relegated to low levels in the Department of the Navy with layering to dilute its influence; (2) be institutionalized so that it cannot easily be disbanded at the whim of a future CNO or SECNAV (the dissolution of the Systems Analysis Division of the Office of the Chief of Naval Operations (OP-96) in the 1980s has long been viewed as a disaster); and (3) be closely linked to program builders. Whether these criteria can be met within the current OPNAV organization was not something that the committee could easily assess in the time available.

The Need for a Small, First-Rate Staff

Particularly for strategic-level analysis, the CNO needs products that can best be obtained by a small, first-rate staff that would include a number of exceptionally talented individuals and future leaders and would be connected well to first-rate outside research-and-analysis organizations. The ideal staff should be seen not just as a collection of operations research personnel, but as a multidisciplinary group with a mix of warriors, policy analysts, systems analysts, engineers, economists, and managers (perhaps with master's degrees in business administration). Further, this staff should include members with outstanding potential and promise (e.g., military who will reach flag rank).

Culture

Much is known about the cultural characteristics of first-rate analytical defense organizations, and the Navy can draw upon its own experiences over the years for examples of both good and bad practice. Generically, however, the good characteristics include the following:

- The ethic of getting the problem straight, even if it revisits guidance or assumptions;
- Loyalty to the boss—the CNO—but also to the Secretary of Defense, the President, and the nation, rather than to Navy warfare areas, platforms, and so on;
- Integrity;
- The mind-set to think joint, but also having the ability to do superb competitive innovation and analysis for the Department of the Navy;
- The mind-set to seek broad, complete analyses rather than analyses to support a superior's talking points;
- Respect and energetic search for empirical and expert information, whether it is obtained from people in the field, through experiments, by augmentation of staff, or from other mechanisms;
- Rigor in everything (but not always in numbers or precision);
- A good process that includes (not always linearly: problem definition, identification of assumptions, a plan for analysis, appropriate tools, and so on);
- A very high ratio of thinking and smart, simple analysis to model running and data analysis (but with subcontracts to specialists);[22]

[22]For many years the OSD's Office of Systems Analysis, later the Office of Programming Analysis and Evaluation, did not use any large and complex models, believing that it was instead essential to remain focused on higher-level issues and relatively reductionist (but not naive) analysis. Today, most analytical shops appear to be tilted far to the big-model extreme, to their detriment.

- Dynamism, such as can be achieved by reasonable turnover rates and constant contact with "outsiders" and their ideas, whether elsewhere in the organization or in Federally Funded Research and Development Centers, professional societies, universities, or industry; and
- The opportunity for self-motivated work that attempts to look beyond the current in-basket and conventional wisdom.

These characteristics can only be developed and sustained within a top-notch analytic organization if the leaders (i.e., the CNO and other key leaders in the Navy) instill and support them.

Links to the Department of Defense and the Other Services

For the Navy, it is important that its analytical shop(s) have good links within the DOD and particularly to the other Services, not just in required coordination meetings but also as a standard part of doing good, professional work on behalf of the nation.

Creating an Appropriate Analytic Organization

With the material provided in earlier subsections as background, the next question is how to go about creating the appropriate analytic organization. The committee's basic recommendation on this question is presented below, followed by a discussion of possible models for the Navy to use in addressing the issue.

Recommendation 3: The Chief of Naval Operations and the Secretary of the Navy should develop a clearly delineated concept of the Navy's future senior-level analytic support organization and define goals for its composition, including multidisciplinary orientation and officers appropriate for high positions.

Potential Models

At least two good models exist for creating and maintaining a highly competent organization to perform analysis and package options for choice. One is similar to the Navy's OP-96 model of the late 1960s and early 1970s and the Army's Office of the Assistant Vice Chief of Staff (AVCS) in the same period. In these cases the organization in charge of the analysis also prepared the resource-allocation decision packages.

The other model was developed by the Air Force. In this case, the position of Assistant Chief of Staff for Air Force Studies and Analyses (AF/SA) was created, with a responsibility limited to performing in-house, independent analyses on issues impacting Air Force operations and programs, current and future. The key distinction was that OP-96 was part of the OPNAV organization responsible for providing resource-allocation packages to the CNO, whereas the AF/SA organization's only function was to do studies and analyses for the Chief of Staff, the Service Secretary, and other Air Force organizational elements requesting analyses (i.e., it did not have a direct resource-allocation responsibility for preparing Air Force programs and/or budgets).

An advantage to having the strong analytic arm working for (or with) those preparing resource-allocation decision packages is that the analysts are grounded in the reality of the issues and types of decisions that must be made each year. However, a potential disadvantage of a direct tie to the resource-allocation staff is that the analytic staff members can get so caught up in the immediate issues that they are not provided the opportunity to build analytic capital and focus on larger, longer-term issues that may be much more important to the future of the Service (a matter of analysis production with available tools versus long-term analytic development with the anticipation of essential issues). In addition, some view a very strong analytic arm in the resource-allocation organization to be an overconcentration of power that may be misused and abused and thus be a detriment to the Service. This source of contention had developed in the Army, and the AVCS office was dissolved, in part because of criticisms regarding its power and effectiveness.

When at its peak effectiveness, the AF/SA office was viewed more as an honest broker of studies and analyses, with independence from functional program responsibilities, that would provide quality products to the Air Force staff and major commands. Those elements would then use the products in conjunction with their own work to prepare decision packages and advocacy positions for programs. The decision maker (e.g., Chief of Staff or Service Secretary) could get a "second opinion" on the decision packages if needed, by asking the Assistant Chief for AF/SA if AF/SA products were being misused. If the AF/SA team had not provided analyses on the issue, it could be asked to provide an independent assessment. A current analogy would be a request by the DOD or a military department acquisition authority for an independent cost estimate of an important acquisition program, or a request for an independent review of the program manager's cost estimate by the Cost Analysis Improvement Group or its Service equivalent.

A key criterion for success is that the CNO and the SECNAV get as close as possible to unbiased analysis and presentation of decision packages. Additionally, they need to have the means to get an independent assessment on important issues when needed. It is best to find out if there are any weak links in a package and to deal with those weaknesses before making a decision on, or recommendation for, a major commitment.

What model would be best for the Navy in the current environment is not clear to this committee at the present time (and, as with much capabilities-based planning, there may be more than one viable solution). The Navy could address the issue in a subsequent tasking to an outside entity, it might do an in-house assessment, or it could make some decision based on available information. This committee believes that the CNO and SECNAV should expend the necessary time and effort to be assured that they are making the best choice for the Department of the Navy. The way in which they choose to obtain and organize their high-level decision preparation and analytic support could be the most important decision that they make regarding possible future success of the Department of the Navy's efforts in capabilities-based planning and overall resource-allocation for a decade or more.

In addition to the analytic organization, the Navy also will need to develop an increased supply of top-flight officers suitable for working in this organization and then moving on to operational assignments and flag rank. Developing this supply of officers will require looking at potential major changes of guidance and process in the personnel system, as well as related changes of the incentive structure.[23] The following is worth noting:

- In the 1960s and 1970s, the Navy made a concerted effort to retour officers in a subspecialty ashore (such as in the planning, programming, and budgeting system (PPBS)), to improve their experience and qualifications in that area. A number of admirals came out of that program and served the Navy and several CNOs well with their experience and insights.
- In the 1980s, the Navy changed its policy and programmed officers with no prior experience into critical PPBS positions—a policy that continues today. The Navy would never consider assigning a captain to command at sea without extensive experience and proven performance in prior sea tours. However, it quite often assigns captains to critical billets in PPBS—on the OPNAV staff, SECNAV staff, and the Joint staff—with little or no prior analytical and resource-allocation (e.g., PPBS) experience or training. Having a small, qualified analytic staff for the Navy will be very difficult to achieve, at least in the near term, until sufficiently experienced officers are developed.

The committee believes that the Navy needs to change some current manpower and personnel policies in order to enhance its ability to build a longer-term, high-quality OPNAV staff with enhanced potential for performing excellent capabilities-based planning and analysis. A key element of those changes should involve creating assignment patterns for future leaders to introduce such individuals early to the discipline of analytical thinking in a real-world context (e.g., the analysis for, preparation of, and review of the Navy Program Objective Memorandum and/or equivalent parts of the overall DOD program). Such assign-

[23]In the past, some distinguished Navy four-star admirals have had Ph.D.s in "hard" disciplines as well as tours in Navy or DOD analysis.

ment patterns should continue to expose such individuals in their careers to the world of analysis and trade-offs in which outcomes influence budgets and/or major programs.

The Temporal Issue: Near Term and Longer Term

There clearly exists a temporal issue with regard to obtaining the type of analytical support that the committee believes the top-level Navy decision makers should have. Actually obtaining such support would take time even if a decision to do so were made immediately. Thus, the committee suggests the following:

Recommendation 4: In the short term, the Chief of Naval Operations and the Secretary of the Navy should go outside their organizations to sharpen concepts and requirements, drawing on the external community of expert practitioners in analysis. Also, they should augment their in-house analytical capabilities in the short term by drawing on Intergovernmental Personnel Act assignments (and other individuals who could take leave from their home organizations), Federally Funded Research and Development Centers, and national and other nonprofit laboratories.

4

Navy Participation in Capabilities-Based Planning Processes of the OSD and OJCS

This chapter contains a description of the status of the capabilities-based planning (CBP) processes of the Office of the Secretary of Defense (OSD) and the Office of the Joint Chiefs of Staff (OJCS). It also provides an assessment of the Navy's participation in these CBP processes and presents recommendations for improvements in the Navy's participation.

HIGHLIGHTS

> I expect all stakeholders in the Department to participate in these efforts to address joint operational needs effectively and to improve the management of Defense Resources.
>
> Donald Rumsfeld, Secretary of Defense
> Memorandum of October 31, 2003

The committee was quick to recognize that the Department of Defense's (DOD's) capabilities-based planning process is not yet well defined or structured for managing defense resources. Nevertheless, the Secretary of Defense is currently using it as his primary process to guide and assess Service program proposals. In that regard, it is important for the Navy to assume a strong role both in shaping the details of the DOD process as it evolves and in structuring Navy program proposals in ways that are responsive to guidance from the Secretary of Defense.

Unfortunately, the information provided to this committee indicates that the Navy is not well represented in the DOD joint CBP processes. Several briefers noted that Navy representatives on the various boards and committees are too often unfamiliar with the issues, or not empowered to speak on behalf of the Navy, or absent. Failure of the Navy to provide adequate representation to the various joint capabilities planning activities clearly complicates or precludes full consideration (and resourcing) of the Navy's potential contributions to joint warfighting. The only visible exception to this problem was seen as being the strong role played by naval personnel from the Pacific Command in the Operational Availability studies that the department is using to help assess joint needs.

BACKGROUND

For managing resources, the DOD uses two types of planning that are often confused, even though they are substantively different. One type is operational planning and the second is future force/program planning. *Operational planning* is the process by which operational commanders at all levels think through, write down, refine, and promulgate to their subordinate commands the courses of action that they intend to be followed in any number of possible situations requiring the use of the military capabilities currently available. This category includes not only such deliberate planning, but also crisis-response adaptive planning. The resulting operational plans form the basis for structuring analyses of the adequacy of current capabilities and for identifying near-term risks and excesses. Such near-term assessments of operational plans are informative and may drive near-term adjustments to existing force and/or equipment allocations and peacetime positioning. As such, they are considered an important element of capabilities-based planning. It is these types of near-term adjustments for which the combatant commanders are the lead customers and proponents within the resource-allocation and management process.

However, current OSD efforts to refine the DOD's emerging capabilities-based planning process focus more on the planning and programming of future forces than on fine-tuning the capabilities of current forces. *Future force/program planning* is the process by which the DOD builds its biennial budget proposals for the funding of future capabilities. The DOD has long been a leader in the employment of a hierarchical, sequential planning, programming, and budgeting system (PPBS; now, planning, programming, budgeting, and execution system (PPBES)). This resource management system builds, revises, and maintains a coherent, visible, multiyear plan for acquiring the military technology, forces, equipment, and infrastructure (including logistic support) appropriate for meeting the future challenges foreseen by the department's leadership. It is this process that is being shifted from having a narrowly threat-based focus to a broader "capabilities-based" focus, as described in Chapter 2. It is particularly important to recognize that the "options" being defined, analyzed, and ultimately chosen in the emerging CBP process are ultimately budgetary choices that must be packaged and explained to the Congress in budgetary terms.

In principle, the DOD's budgetary choices reflect the following steps:

- Start with the administration's strategic goals;
- Include explicit economic constraints on the DOD;
- Define appropriate Strategic Planning Guidance, together with suitably broad (in the parametric sense described in Chapter 2) Defense Planning Scenarios;
- Solicit Service program/budget proposals (Program Objective Memorandums (POMs)) that are responsive to this guidance; and
- Direct key analytic studies that develop and assess options both within and across the Services.

Input is also solicited from the current combatant commanders, but in this process the Secretary of Defense acts, in effect, as a surrogate for future combatant commanders in that he commands the tools needed to best estimate future threats and potential future U.S. capabilities, including those driven by new technology. The latest revision to this process is depicted in Figure 4.1.

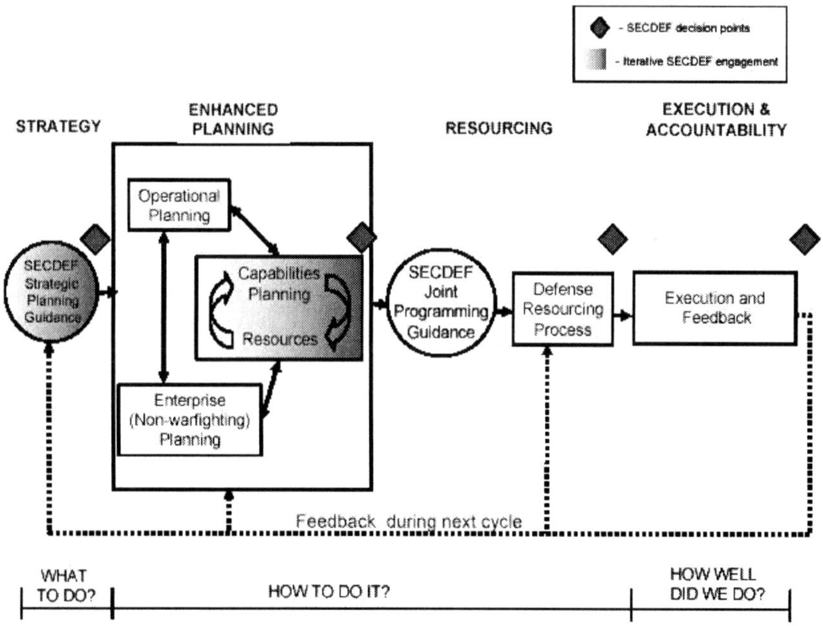

FIGURE 4.1 The new process for the Department of Defense's future force/program planning. NOTE: SecDef, Secretary of Defense. SOURCE: Joint Defense Capabilities Study Team. 2004. *Joint Defense Capabilities Study: Improving DoD Strategic Planning, Resourcing, and Execution to Satisfy Joint Requirements,* Final Report, Department of Defense, Washington, D.C., January, p. v.

Of particular note is the intent that the likely limits on available funding will inform the program planning process starting with the initial Strategic Planning Guidance, rather than being disruptively imposed at the end of the process as the next budget request is being finalized. In principle, each Service will no longer be able to argue that it needs more money if it is to follow the Secretary's guidance. The analytic heart of this process is the Enhanced Planning Process and its associated Analytic Agenda, which is designed to identify and assess specific elements of DOD's future capabilities for early decision by the Secretary and promulgation in the Joint Programming Guidance, before the Services build their POMs. An additional aspect of the new process worth highlighting is the increased role of the combatant commanders, who are now participating much earlier in the planning cycle than was previously the case. In this regard, they are encouraged to express their needs in terms of "capabilities," not specific budgetary items.

Weapons Systems Requirements and Acquisition Management

In addition to the primary planning, programming, budgeting, and execution (PPBE) process, the DOD has also historically employed two other management systems that impinge on the capabilities-based planning process—the weapons systems requirements process and the acquisition management system.

Weapons Systems Requirements

The weapons systems requirements process has traditionally been the province of the military Services. The respective Services are the experts in their forms of warfare and have routinely been the initiators of formal "requirements" for new weapons systems (including command, control, communications, computers, intelligence, surveillance, and reconnaissance (C4ISR) systems). These formal requirements in turn are refined as performance capabilities requirements suitable for inclusion in requests for proposals and subsequent contracts.

The growing emphasis on the need for more "jointness" in military operations has also led to efforts to increase jointness among the Services in their acquisition of weapons systems—particularly those important to battlefield interoperability among Service elements. The resulting Joint Requirements Oversight Council (JROC) review of major new, Service-proposed weapons systems was intended to support this goal. However, the JROC has usually been viewed as only a minor bureaucratic speed bump that might slow but not alter Service programs. It has been viewed this way in part because most programs are well advanced by the time they are reviewed by the JROC, in part because there has been little truly *joint* warfighting doctrine or planning that could inform critical review, and in part because there is little incentive for military officers on the Joint Staff to reach conclusions at odds with those of their parent Service. Fur-

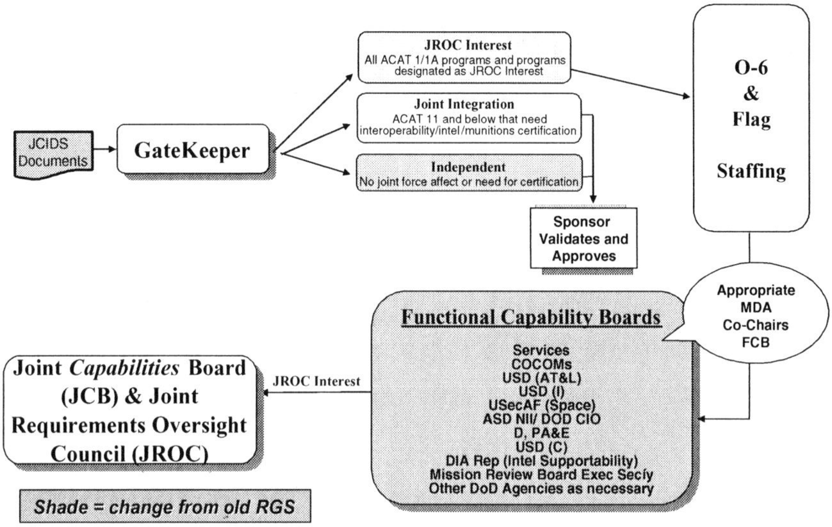

FIGURE 4.2 The Joint Capabilities Integration and Development System (JCIDS) process. SOURCE: LTC Robert Larsen, USA, Chief, Plans Branch, Joint Capabilities Development, and J8, Director for Force Structure, Resources, and Assessment, "As Warfighting Capability Based Analysis and Assessment Evolves: Are We Solving the Right Problem?," presentation to the committee, July 27, 2004, Woods Hole, Mass., slide 42.

thermore, the fact that the JROC only reviews major weapons systems ensures that most of the command, control, and communications and the intelligence, surveillance, and reconnaissance programs that are important to mechanizing true jointness are not being formally considered in the joint arena.

One of the transformational initiatives of Secretary Rumsfeld was to increase emphasis on the joint nature of military operational and future planning, which includes emphasis on ensuring that new weapons systems are "born joint."[1] This latter emphasis results from repeated military interoperability problems in the field over the past several decades, despite repeated policy directives intended to preclude such problems. In 2003 this initiative resulted in the Chairman of the Joint Chiefs of Staff's (CJCS's) formally changing the Joint Staff requirements process by establishing the Joint Capabilities Integration and Development System (JCIDS). This process is illustrated in Figure 4.2.

---

[1]DOD 2003. Transformation Planning Guidance. Available at <http://64.233.161.104/search?q=cache:tFLupyoy9wkJ:www.oft.osd.mil/library/library_files/document_129_Transformation_Planning_Guidance_April_2003_1.pdf+DoD+transformation+&hl=en>. Last accessed on December 3, 2004.

The responsibilities of the the new JCIDS are to provide an enhanced methodology utilizing joint concepts that will (1) identify and describe existing or future shortcomings, as identified against current or future capabilities or as measured against current or projected threat capabilities; (2) identify and describe redundancies in warfighting capabilities; (3) describe the attributes of effective solutions; and (4) identify the most effective approach or combination of approaches to resolve those shortcomings.

The JCIDS process could be interpreted in some ways as intending to deal with elements of the DOD force structure and posture beyond just weapons systems. As a practical matter, however, JCIDS and its several functional components are focused almost exclusively on the acquisition of new equipment and not on force levels or postures.

Acquisition Management

Once a formal requirement for a new weapon system has been established (and approved by the JROC for major new systems) and a Service has included funds in its future-years program proposal, subsequent development and acquisition are governed by the formal DOD acquisition process. This process is substantively separated from the PPBES but is ultimately subordinate to it in the sense that decisions in the PPBES (in the DOD and in Congress) determine the flow of funds to each individual weapons system program.

Each of the past two administrations repeatedly attempted to revise the formal weapons systems acquisition process to remove perceived barriers to the acquisition of cutting-edge commercial technology and also to speed up the fielding of new systems. Despite extensive rewriting of procedural documents and some improvement in the department's ability to acquire commercial products, the sound systems-engineering practices recommended by the Packard Commission almost two decades ago remain in place: Fly-before-Buy, do milestone reviews, and do careful cost-performance tracking of progress.[2] Furthermore, many of the perceived impediments to the use of speedier procurement practices more like those in the commerical world are rooted in congressional requirements for full and open competition, detailed reporting of costs, frequent outside audits, and other steps designed to protect the integrity of the use of taxpayer funds. These considerations, which are firmly embedded in the culture of the DOD acquisition community, have tended to stymie many aspects of proposed acquisition reforms.

---

[2]Report of the President's Blue Ribbon Commission on Defense (also known as the Packard Commission). 1986. Available at <http://www.ndu.edu/library/pbrc/pbrc.html>. Last accessed on December 3, 2004.

Although DOD acquisition management has, at least in recent years, focused on overseeing programs individually, there has been a recent recurrence of interest in broader, "capabilities area" analysis within the acquisition community. This approach is depicted in the Figure 4.3.

The roadmaps depicted at the intersection of the requirements phase and the acquisition phase in this figure are intended to result from "capabilities area" studies to be led by the acquisition community. To the committee's knowledge, this aspect of the DOD capabilities-based planning process has yet to be reconciled with the JCIDS and PPBE efforts.

Recent Activity

The *Joint Defense Capabilities Study* (also known as the Aldridge Study) is a major independent study chartered by the Secretary of Defense "to examine how the Department of Defense (DoD) develops, resources, and provides for joint capabilities."[3] The study group recommended "a capabilities-based process for identifying needs, creating choices, and providing capabilities," and it attempted to address capabilities-based planning across all elements of the future defense program.[4] One member of the present committee was assigned to that study throughout its term. The OSD's Program Analysis and Evaluation official who briefed the committee was also a member of the Aldridge Study team.

The Aldridge Study was favorably received by the Secretary, who directed that its major recommendations be implemented, except for organizational changes. At the time of this committee's workshop, it was clear that the implementation of the study's capabilities-based planning recommendations was at an early stage, with many conceptual and procedural details remaining to be worked out. It is this ongoing implementation process that provides the Navy with an opportunity to contribute to and influence the resulting new resource management process.

An overview of the capabilities-based planning process recommended by the Aldridge Study is presented in Figure 4.4.

As may be seen, it is the Secretary's intent that the department's program planning process become more output- and capabilities-focused, and that major options be developed, analyzed, and brought forward to him for decision. Unfor-

---

[3]Joint Defense Capabilities Study Team. 2004. *Joint Defense Capabilities Study: Improving DoD Strategic Planning, Resourcing, and Execution to Satisfy Joint Requirements*, Final Report, Department of Defense, Washington, D.C., January, p. iii. The study was led by outgoing Under Secretary of Defense (Acquisition, Technology, and Logistics), E.C. "Pete" Aldridge, Jr.

[4]Joint Defense Capabilities Study Team. 2004. *Joint Defense Capabilities Study: Improving DoD Strategic Planning, Resourcing, and Execution to Satisfy Joint Requirements*, Final Report, Department of Defense, Washington, D.C., January, p. iv.

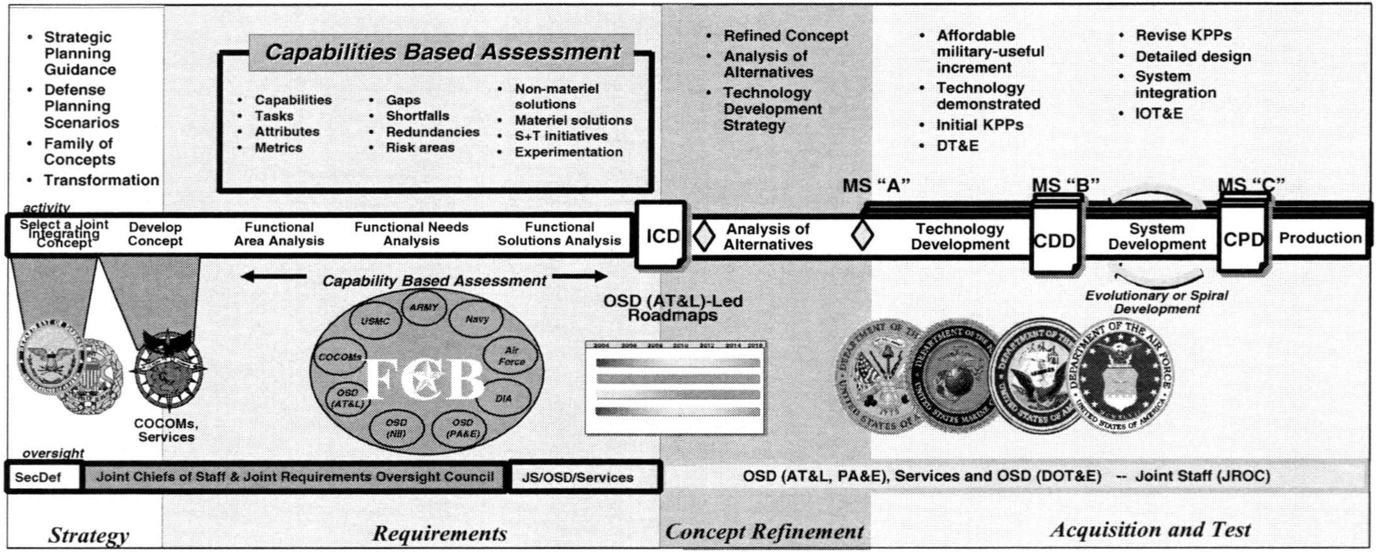

FIGURE 4.3 The end-to-end requirements, acquisition, and test process of the Department of Defense. SOURCE: Strategic Planning Guide study (Acquisition, Technology, and Logistics) Transformed Capabilities Assessment (draft).

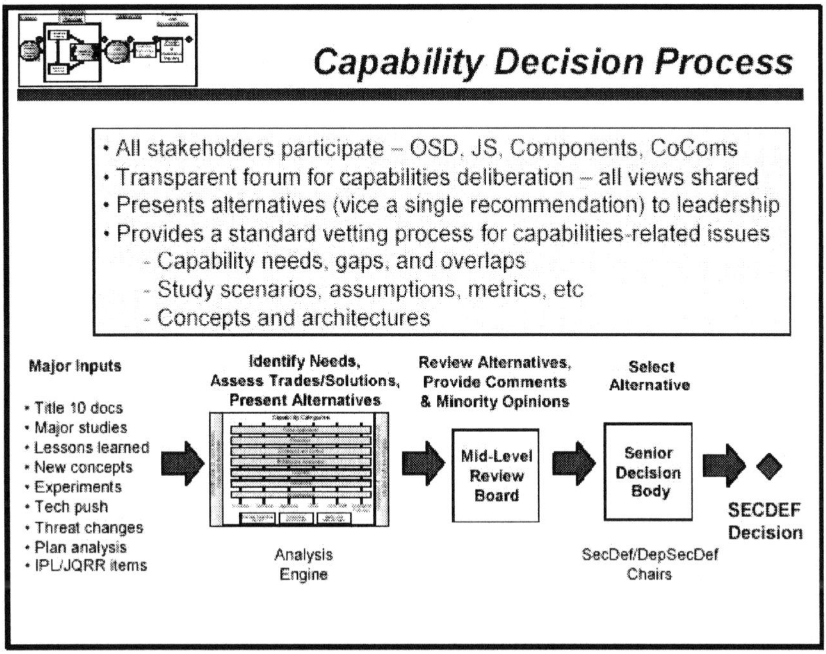

FIGURE 4.4 Capability decision process as recommended by the Aldridge Study. NOTE: OSD, Office of the Secretary of Defense; JS, Joint Staff; CoComs, combatant commanders; IPL/JQRR, Integrated Priority List/Joint Quarterly Readiness Review; SECDEF, Secretary of Defense. SOURCE: Joint Defense Capabilities Study Team. 2004. *Joint Defense Capabilities Study: Improving DoD Strategic Planning, Resourcing, and Execution to Satisfy Joint Requirements,* Final Report, Department of Defense, Washington, D.C., January, p. B-17.

tunately, the Aldridge Study was not able to describe in useful detail the process intended for mapping the broad capabilities categories defined in the study[5] into the programmatic and budgetary detail necessary for constructing and assessing clear budgetary alternatives for the Secretary's decisions. As a result, there continues to be an extensive follow-on staff effort involving OSD, the Joint Staff, and all Services and agencies led by J8/USD (P)/PA&E (Director for Force Structure, Resources, and Assessment/Under Secretary of Defense for Policy/

[5]Joint Defense Capabilities Study Team. 2004. *Joint Defense Capabilities Study: Improving DoD Strategic Planning, Resourcing, and Execution to Satisfy Joint Requirements,* Final Report, Department of Defense, Washington, D.C., January.

Program Analysis and Evaluation). The committee understood from both Navy and non-Navy briefers that the Navy's involvement in these ongoing efforts is marginal, which in turn sharply limits the Navy's ability to influence the emerging process.

The goal of the ongoing joint effort is to refine the capabilities-based planning process to allow the Secretary's intent to be more fully realized. This staff effort involves a multiplicity of Joint Staff/OSD/Service activities that are intended to bring more substance and coherence to the new JCIDS, acquisition, and PPBES processes that are to be integrated into a cohesive whole. These staff activities include the drafting and coordination of documents such as formal Joint Operational Concepts, Joint Integrating Concepts, and Joint Operating Concepts. In addition, the capabilities categories initially nominated by the Aldridge Study have had to be partitioned into more manageable aggregations of capabilities suitable for analysis and programmatic decisions. This process involves new Functional Area Analyses, Functional Needs Analyses, and Functional Solutions Analyses that have the potential, at least in theory, of developing real alternatives to programs preferred by the Services.

NAVY PROGRAM PLANNING

It is the Navy's approach to the capabilities-based aspects of the DOD program planning process described above that has been a major focus of this committee's deliberations. The CNO's guidance to the Navy for 2004 emphasized joint capabilities. However, the committee saw very little evidence of jointness in the process that the Navy uses to develop its multiyear program plan, as that process was explained to the committee. Furthermore, the committee saw no evidence that the CNO's guidance was updated in response to the Secretary of Defense's Strategic Planning Guidance, which was issued late in the cycle.

The Navy's choice of its Sea Power 21 construct as the primary basis for planning its forces is a potential source of a significant problem. There is not yet adequate linkage between the Sea Power 21 framework and the DOD capabilities-based planning framework. This dichotomy is recognized by the Navy, as illustrated in Figure 4.5.

Supporting Analysis

The Navy has a long history of conducting detailed quantitative analysis at several levels of aggregation, in support of both Navy program planning and the operational planning of naval component commanders. The committee was briefed in general on several of the Navy's ongoing analytic efforts, including major campaign analyses of potential future conflicts. Such studies were conducted at several locations within the Navy staff, its field activities, and system commands. Principal offices include the Deputy Chief of Naval Operations

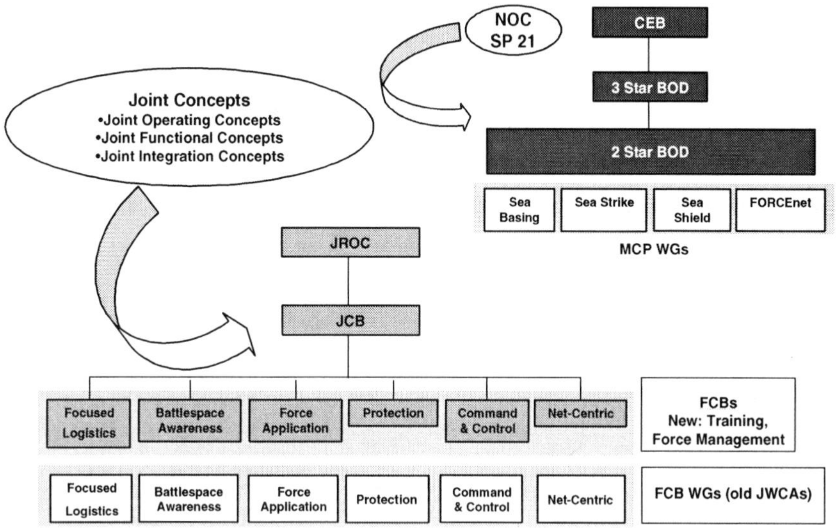

FIGURE 4.5 Navy and Joint Capabilities Integration and Development System (JCIDS) executive and decision-making process. NOTES: NOC, Naval Operating Concept; SP 21, Sea Power 21; CEB, CNO Executive Board; BOD, Board of Directors; MCP WG, Mission Capability Planning Work Group; JROC, Joint Requirements Oversight Council; JCB, Joint Capabilities Board; FCB, Force Capabilities Board; JWCA, Joint Warfighting Capabilities Assessment. SOURCE: LCDR Kenneth M. Masson, USN, Office of the Deputy Chief of Naval Operations for Resources, Requirements, and Assessments, N815F/N00X, "Capabilities Based Planning—Integrating into the Joint Process," presentation to the committee, July 27, 2004, Woods Hole, Mass., slide 10.

(DCNO) for Warfare Requirements and Programs (N7), which is responsible for the end-to-end building and integration of the Navy Future Force Plan/Program Objective Memorandum (POM); the Assessments Division of the Office of the DCNO for Resources, Requirements, and Assessments (N81), which is responsible for assessing the POM on behalf of the CNO and interfacing with the joint/OSD future force planning process; and the systems commands that are responsible for the architectures and interoperability assessments of each of the pillars, or major naval concepts, of Sea Power 21. Additional campaign-type analyses are conducted by the Navy Warfare Development Command (NWDC), but these are primarily in support of near-term operational planning needs, not longer-term force planning.

The material presented to the committee indicated that many of these Navy campaign analyses are structured around scenarios that the Navy believes are appropriate for its purposes. These scenarios do not, however, comport in significant details with the Defense Planning Scenarios that the Secretary's office has promulgated for use department-wide in all future force planning. It is understandable that the Navy would want to conduct excursions from the basic analytic framework used elsewhere in the department. Nonetheless, the committee believes that the Navy's credibility with OSD is jeopardized by its not clearly employing these DOD scenarios as basic building blocks in its force planning process. The fact that N81 may come close to employing these scenarios in its assessment process at the end of the POM development cycle does not compensate adequately for their lower priority during the formative stages of the POMs. Indeed, the committee observes that the analytic capabilities of N81 appear insufficiently robust to compensate effectively for their basic shortcoming. In particular, the N81 analyses briefed to the committee were narrowly focused on specific issues and did not purport to critically review or replicate the overall POM development process as conducted within the N7 organization. Indeed, much of the analytic capability within the Navy staff appears to be excessively stovepiped and unresponsive to the higher needs of capabilities-based future force planning. Only in the case of the Office of the DCNO for Fleet Readiness and Logistics (N4) and, to a lesser extent, the Office of the DCNO for Manpower and Personnel (N1) did it appear to the committee that analytic efforts were directly responsive to top-level guidance.

More generally, the Navy's force planning analytic activities were not shown to the committee to be linked in any meaningful way with the Aldridge Study's Enhanced Planning Process and its associated Analytic Agenda, endorsed by the Secretary of Defense in October 2003.

An Example: Homeland Defense

Homeland defense—that is, the direct defense of the United States against external attack—is described as the DOD's highest priority. Yet, with the exception of the Navy's program to grow the capability of its Aegis ships to detect, track, and eventually engage a larger spectrum of cruise and ballistic missiles, there was no evidence presented to the committee of Navy programs that are intended, for example, to provide broad, active surveillance of the approaches to the United States from which terrorist or hostile state attacks could be launched in the future. While the committee understands that the capabilities needed to accomplish this mission have not been fully defined, the fact that the mission was nowhere visible in the material presented to the committee was troubling.

Navy/Marine Corps "Jointness"

A significant portion of the U.S. Navy is directly committed to providing effective joint capabilities with the U.S. Marine Corps. This is increasingly true as the Navy has shifted its emphasis to operations in the littorals, including greater emphasis on sea basing. However, in its workshop, the committee was shown no evidence of the type of integrated Navy/Marine Corps program planning and analyses that one would expect if the Department of the Navy was attempting to optimize its investment in such combined forces. This suggests that the Navy and Marine Corps are continuing to develop their individual programs in excessive isolation. The Navy's ability to interact productively in the larger program planning process involving joint forces capabilities could well be enhanced by being able to present a program well-integrated with that of the Marine Corps. This program could be a model for future use as the need to jointly plan future sea basing with the Army and Air Force evolves.

RECOMMENDATIONS

Recommendation 5: The Chief of Naval Operations should direct that Navy force planning consistently include use of the baseline scenarios (including concepts of operations, threat assessments, and so on) specified by the Office of the Secretary of Defense. The Chief of Naval Operations should also work with the Secretary of the Navy and the Commandant of the Marine Corps to establish a more integrated, joint Navy/Marine Corps Program Objective Memorandum development process that could serve as a model for the Department of Defense more broadly. Navy force planning should also include extensive exploration to lead to better understanding of the consequences of scenario details and other assumptions of analysis.

Of course, within the extensive exploration recommended above, Navy planning should address in particular depth those scenarios and the cases within them that affect Navy missions, capabilities, and needs. The point here, however, is that the Navy should address the DOD's standard baseline cases to be a creditable joint participant.

During the workshop, Navy briefers described their organizations' attempts to cover all the bases in CBP meetings called by OSD and OJCS personnel, but indicated that the organizations had insufficient resources to participate fully and effectively. Some lack of coordination and of internal Navy scheduling control was also evident. This may be due in part to the lack of clarity in OPNAV lines of responsibility for CBP activities; that lack sometimes results in a duplication of effort. In addition, other information provided to the committee indicates that the Navy is not always well represented in the DOD joint CBP processes. Several non-Navy briefers noted that Navy representatives on the various boards and committees are often unfamiliar with the issues, not empowered to speak on behalf of the Navy, or absent. Regardless of whether these difficulties result from insufficient staff resources or other problems, they preclude justice being done to naval issues in the larger DOD environment and undercut Navy interests in competition for both influence and funding.

Recommendation 6: Over the longer term, the Chief of Naval Operations should identify and staff a central activity that is charged with the responsibility for harmonizing the Navy's capabilities-based planning processes with those of the larger Department of Defense (DOD) force planning community, and should staff that activity to represent Navy interests effectively in the DOD joint planning activities.

Currently, the most logical place for this assignment would be in the Office of the DCNO for Resources, Requirements, and Assessments, because the processes involved in capabilities-based planning include resource-allocation issues across all force and support areas. However, the decision on how best to accomplish the tasking for such a central activity should logically follow from or be consistent with the model chosen by the CNO to assure having a highly competent analytical group to support him on all aspects of capabilities-based planning. In the short term, to make the best possible utilization of resources, a possible solution would be to designate one individual (e.g., the director of the Navy staff) to be responsible for resolving any capabilities-based planning coordination problems within the OPNAV staff.

5

Potential Future Efforts

This chapter addresses additional inputs and assessment areas that could not be covered within the committee's limited work plan and that could be of benefit to the Navy in improving its analytical activities and capabilities-based planning (CBP).

ADDITIONAL INPUTS

The committee's recommendations in this report are based on its workshop briefings, on the combined experience and backgrounds of the individual committee members, and on additional documentation (e.g., the *Joint Defense Capabilities Study*—called the Aldridge Study[1]) reviewed by the committee. The rapid start-up of the committee and the limited time available for it to gather and analyze information and to complete this report did not allow for input from many potentially important sources and on certain important topics, including the following:

- *U.S. Marine Corps*—This report addresses the Navy's activities on the basis of the tasking to the committee. In the committee's brief workshop, held July 27-29, 2004, no input was received from the Marine Corps about its implementation of capabilities-based planning and the mechanism by which its process

[1]Joint Defense Capabilities Study Team. 2004. *Joint Defense Capabilities Study: Improving DoD Strategic Planning, Resourcing, and Execution to Satisfy Joint Requirements*, Final Report, January, Department of Defense, Washington, D.C.

is linked to that of the Office of the Chief of Naval Operations (OPNAV), despite the joint flavor of the 2004 guidance from the Chief of Naval Operations (CNO), which includes the words "Partner with the Marine Corps to identify and develop enablers to achieve Ship-to-Objective Maneuver (STOM)."[2]

- *Combatant commanders*—The future force planning component of the capabilities-based planning process is intended to resource those capabilities that will be needed by the "tip of the spear" to meet future needs. Time constraints precluded the committee's seeking input from selected combatant commanders (e.g., the Pacific Command) in order to understand the degree to which these commanders engage with the Navy during the CBP process and whether the Navy should interact more with them or their subordinate commanders.
- *U.S. Joint Forces Command*—The U.S. Joint Forces Command is the Department of Defense's (DOD's) lead command for transformation and the natural lead organization in defining the capabilities needed by future combatant commands. As a follow-on task, the Navy could address the degree of connectivity between this command and the Navy within the CBP process and evaluate different aspects of that connectivity, as appropriate.
- *Office of the Secretary of Defense*—Historically, many new and important capabilities (e.g., the Global Positioning System, precision weapons, sea-launched ballistic missiles, stealth, A-10 and F-16 aircraft) have been introduced not by demand pull from combatant commanders or their predecessors (the commanders in chief), but rather from a combination of technology push and strategic planning. The Office of the Secretary of Defense (OSD) (and elements within the Services) continue to have an important role in encouraging such ideas, and the Navy could potentially benefit from more interaction with the Under Secretary for Acquisition, the Director of Transformation, or their representatives.
- *Homeland defense*—A high priority is assigned to homeland defense in the DOD's Planning Guidance. However, the committee was not exposed to much information about the DOD's CBP activities specific to homeland security or homeland defense. Pursuing this area in more depth could benefit the Navy.
- *Non-DOD implementations of CBP*—The committee did not have sufficient time to examine the tools used to implement CBP outside the DOD. The Navy could address the potential application of analytic and other planning tools used by very large corporations to Navy CBP. Examples of potential subjects could include methods of using available but uncertain intelligence information about competitors or threats, and the applicability of commercial planning and analysis tools to Navy functional support areas that have similarities to commercial operations.

---

[2]ADM Vern Clark, USN, Chief of Naval Operations. 2004. *CNO Guidance for 2004*, Department of the Navy, Washington, D.C., p. 15.

ADDITIONAL ASSESSMENT AREAS

Potential areas for further review and assessment that could significantly affect future Navy capabilities-based planning activities include the following:

- A recommendation in this report relates to the need for the CNO to be advised by a small, top-notch group of analysts. The Navy could better characterize alternative structures for this group, including its relative benefits and its needs (e.g., sources of its membership and the types of education and training needed by its members) before making a decision on how it will proceed in this area.
- As noted in Chapter 1, the committee found that most of the Navy's, and the entire DOD's, CBP activities are focused on the resourcing for weapons and platforms and their operations. The committee is convinced that a good CBP process should also address other important resource areas, including human resources, in which personnel policies and education and training practices impact the utilization of Navy personnel. These functional areas consume large portions of Navy funding and directly impact all Navy capabilities—in particular, force readiness. While the committee briefly heard from representatives of the Deputy Chief of Naval Operations (DCNO) for Manpower and Personnel (N1) and the DCNO for Fleet Readiness and Logistics (N4) about human resources and selected readiness analytic efforts, the Navy needs to better integrate their activities into the Navy's capabilities-based planning process.
- In its workshop the committee was unable to address the role of gathering, processing, and disseminating intelligence in achieving capabilities and the use of intelligence information in preparing CBP assessments. The gathering of good intelligence information, the availability of such information to analysts, and their use of it in analyses and assessments are all contributors to estimating the risk associated with different alternatives available to operational commanders and to force planners and programmers. Intelligence is a function often assumed away in many models. It is an important subject for review in Navy CBP and analysis and is beyond the purview of much of the OPNAV staff.
- After its July 2004 workshop, the committee learned that the Assessments Division of the Office of the DCNO for Resources, Requirements, and Assessments (N81) is funding the integration of the Naval Simulation System (NSS) with the U.S. Army's and the U.S. Marine Corps's Combat XXI activities.[3] This effort could provide a useful simulation tool to support the Department of the Navy's CBP process, especially as it pertains to joint capabilities involving the

[3]This project is entitled Office of the Chief of Naval Operations, N81, and the Naval Postgraduate School World-Class Modeling (WCM) Project.

Army and Marines. The Navy could benefit from a review of this integration effort to develop a good understanding of the scope and deliverables and, if appropriate, use the results to improve its own CBP process.

- A new concept being floated within the DOD currently involves "capability deployment groups." The motivation for the concept stems from the recognition that implementation of grand concepts for capability development will not be realized in the absence of sustained and focused implementation mechanisms. The Navy could assess this concept and its potential to improve Navy CBP activities.

Appendixes

A

Committee and Staff Biographies

John D. Christie (Chair) is a senior fellow at LMI. He has an extensive background in Department of Defense (DOD) resource allocation, acquisition policy, and program analysis. From 1989 to 1993, Dr. Christie was the Director of Acquisition Policy and Program Integration for the Office of the Under Secretary of Defense (Acquisition). In this role, he directed the preparation of a comprehensive revision of all defense acquisition policies and procedures and prepared comprehensive acquisition program alternatives for the Secretary of Defense that resulted in multibillion-dollar budget reductions. From 1972 to 1976, Dr. Christie served as the Principal Deputy Assistant Secretary of Defense for Program Analysis and Evaluation. As a member of Army Science Board in the 1980s, he was called upon to direct reviews of the Army analytical community and operations research activities for the Vice Chief of Staff, including the support of the overall Army acquisition process and its integration with the programming and budgeting process. From 1994 to 1995, Dr. Christie coordinated a process team effort for the Commission on Roles and Missions of the Armed Forces; this team provided the commission with recommendations for improving defense management of the DOD planning processes with respect to requirements generation and resource allocation. Dr. Christie has also been an active participant in studies of the National Research Council (NRC), having served as chair of the committee that produced *Recapitalizing the Navy: A Strategy for Managing the Infrastructure* (1998), and as a member of several other NRC study committees, most recently the Panel on Survivability and Lethality Analysis and the Committee for the Role of Experimentation in Building Future Naval Forces. He is a member of the NRC's Naval Studies Board.

Paul K. Davis is senior scientist and research leader at RAND, where he works primarily on defense planning. He is also a professor on the faculty of the Pardee RAND Graduate School. He served a 5-year tour at RAND as corporate research manager for defense and technology planning and was program manager for strategy planning and assessment. Prior to joining RAND, Dr. Davis was a senior executive in the Office of the Secretary of Defense for Program Analysis and Evaluation (OSD/PA&E). He holds a B.S. from the University of Michigan and a Ph.D. from the Massachusetts Institute of Technology in chemical physics. His most recent monographs involve capabilities-based planning, effects-based operations, and strategy for the deterrence and influence component of counterterrorism.

John F. Egan is an independent consultant, having retired in 1998 as a senior corporate officer and vice president for corporate development at Lockheed Martin Corporation. He has an extensive background in the management of high-technology enterprises involving both the public and private sectors in defense, aerospace, electronics, and information systems. During his 25-year career with Lockheed Martin (including Sanders and Lockheed), Dr. Egan held diverse executive positions responsible for acquisitions, mergers, joint ventures, strategic planning, business development, and research and development. He also was general manager of two separate profit/loss divisions, respectively engaged in design, engineering, manufacture, sales, and support of (1) defense electronics and (2) commercial computer equipment. Dr. Egan also has a broad understanding of defense programs, business and strategic planning, and acquisition and policy. An electrical engineer by training, he is a former chief scientist for the Chief of Naval Operations (CNO), has held senior technical positions in the Office of the Director of Defense Research and Engineering and the Air Force, and has extensive experience with systems for command and control, intelligence, and electronic and information warfare. He is a current member of the CNO's Executive Panel and Chair of the NRC's Naval Studies Board.

Kerrie L. Holley is chief technology officer for IBM's Web services and Service-Oriented Architecture (SOA) Center of Excellence. He is chief architect for the Application Innovation Services unit. His expertise includes translating business requirements into process designs for cutting-edge, network-centric distributed solutions. An IBM Distinguished Engineer and a member of the IBM Academy of Technology, Mr. Holley has focused on issues related to the modernization of legacy networks and databases to take advantage of Web-services-based computing technologies. Currently, his interests include Web services and e-business solutions, including technical oversight, information technology consulting, adaptive enterprise architecture design, information technology strategy, formation of partnerships among clients and vendors, and managment of technical risks. Mr. Holley holds a B.A. degree in mathematics from DePaul University and a J.D. from the DePaul School of Law. He is a member of the NRC's Naval Studies Board.

Samuel D. Kleinman is director of the Infrastructure and Readiness Team at the Center for Naval Analyses (CNA). Dr. Kleinman has an extensive background in the evaluation of business practices, and at CNA he is responsible for infrastructure and financial issues central to the Department of the Navy. His research interests include reducing the infrastructure, base consolidation, outsourcing, housing, management efficiency, industrial base, acquisition reform, material support, and transportation. Dr. Kleinman has been a participant on numerous scientific boards and advisory committees, such as the Defense Science Board's 1996 study of operating and support costs.

R. Bowen Loftin is currently at Texas A&M University at Galveston. He was formerly executive director of the Virginia Modeling, Analysis and Simulation Center and director of simulation programs at Old Dominion University (ODU), with responsibility for the university's graduate programs in modeling and simulation. Dr. Loftin joined ODU in Norfolk, Virginia, in 2000, as professor of electrical and computer engineering and professor of computer science. Previously he had been professor and chair of the Department of Computer Science and director of the National Aeronautics and Space Administration's (NASA's) Virtual Environments Research Institute at the University of Houston. Since 1983 Dr. Loftin, his students, and coworkers have been exploring the application of advanced software technologies, such as artificial intelligence and interactive, three-dimensional computer graphics, to the development of training and visualization systems. He is a frequent consultant to both industry and government in the area of advanced training technologies and scientific and engineering data visualization. Dr. Loftin has served on numerous advisory committees. Awards that he has received include the University of Houston-Downtown Award for Excellence in Teaching and Service, the American Association of Artificial Intelligence Award for an innovative application of artificial intelligence, NASA's Space Act Award, the NASA Public Service Medal, and the 1995 NASA Invention of the Year Award. He is the author or coauthor of more than a hundred technical publications.

L. David Montague (NAE), an independent consultant, is retired president of the Missile Systems Division at Lockheed Martin Missiles and Space and a former officer of Lockheed Corporation. A member of the National Academy of Engineering (NAE), he has more than 40 years of experience in the design, development, and program management of military weapons and their related systems. His expertise includes complex systems engineering and systems integration of ballistic missiles, cruise missiles, and unmanned aerial vehicles. Mr. Montague is a fellow of the American Institute of Aeronautics and Astronautics (AIAA) and a previous recipient of the AIAA's Missile Systems Award. He has served on numerous scientific boards and advisory committees, including the Navy Strategic Systems Steering Task Group and task forces for both the U.S. Army and the Defense Science Board. Mr. Montague is a member of the NRC's Naval Studies Board.

Gene H. Porter is an independent consultant in matters relating to national security planning and weapons system development. His current clients include the Institute of Defense Analyses. Most recently, Mr. Porter has been supporting the Office of the Secretary of Defense (OSD) in defining the detailed Defense Planning Scenarios that are intended to guide the development of U.S. military force posture and modernization programs through the end of the decade. This analytic work has involved an all-source examination of potential threats, including space-based threats, and potential U.S. responses. Previously, Mr. Porter had served as director of Acquisition Policy and Program Integration for the Office of the Under Secretary of Defense for Acquisition, where he was responsible for long-range planning, programming, and budgeting matters related to new military warfare systems. His earlier career included various staff and line management positions at Lockheed-Sanders Corporation in the development and manufacture of military and commercial electronic systems, including mine and undersea warfare systems. Mr. Porter has served on numerous scientific and advisory committees, including as chair of the NRC Committee for Mine Warfare Assessment.

Andrew P. Sage (NAE) is founding dean emeritus of the School of Information Technology and Engineering at George Mason University (GMU) and First American Bank Professor of the GMU Department of Systems Engineering and Operations Research. His research interests include systems engineering and management efforts in a variety of application areas, such as systems integration and architecting, reengineering, software systems engineering, total quality management, cost and effectiveness assessment, and industrial ecology and sustainable development. Prior to joining the faculty at GMU he served on the faculty at the University of Arizona, the University of Florida, Southern Methodist University, and the University of Virginia. Dr. Sage has served on numerous scientific boards and advisory committees. He is a fellow of the Institute of Electrical and Electronics Engineers, the American Association for the Advancement of Science, and the International Council on Systems Engineering, and a member of the National Academy of Engineering. Dr. Sage received a B.S. in electrical engineering from the Citadel, an S.M. in electrical engineering from the Massachusetts Institute of Technology, and a Ph.D. from Purdue University.

Elaine Simmons, currently in the Office of the Secretary of Defense and formerly a research fellow at LMI, is experienced in DOD strategic planning, needs assessment, and joint, collaborative analytic studies. She was the LMI lead on the Joint Defense Capabilities Study (known as the Aldridge Study), which developed a new process and organizational alternatives for making joint needs the foundation for the defense program. She also led a study for the Office of Force Transformation that developed Active/Reserve Component options using a capabilities-based approach. Ms. Simmons came to LMI from the Navy, where, as a member of the Senior Executive Service (SES), she was head of the Campaign Analysis and Modeling Branch in the Assessment Division/N81 in the

Office of the Chief of Naval Operations (OPNAV). Prior to working at OPNAV, she was a staff member in OSD/PA&E, serving as the director of the Joint Data Support organization, which provided data and analytic products to large joint studies such as the Mobility Requirements Study 2005. Before joining the government, Ms. Simmons was an operations director for GRC International. She earned a master's degree in defense policy analysis from the George Washington University and a bachelor's degree in international studies from George Mason University.

William D. Smith retired as an admiral in the U.S. Navy after 38 years of active-duty service. Admiral Smith's background is in Navy planning, programming, budgeting, and operational issues. His last assignment was as U.S. Military Representative to the NATO Military Committee in Brussels, Belgium. Admiral Smith has served in a number of high-ranking capacities for the Chief of Naval Operations. From 1987 to 1991, he served as Deputy Chief of Naval Operations for Logistics and Navy Program Planning. From 1985 to 1987, he was director of the Fiscal Management Division/Comptroller of the Navy. Admiral Smith is a member of the NRC's Naval Studies Board.

Staff

Charles F. Draper is director of the National Research Council's Naval Studies Board. He joined the NRC in 1997 as a program officer of the Naval Studies Board, later was promoted to senior program officer, and in 2003 became associate director of the board. During his tenure with the board, Dr. Draper has served as the responsible staff officer on a wide range of studies aimed at helping the Department of the Navy with its scientific, technical, and strategic planning. His recent efforts include studies in the areas of network-centric operations, theater missile defense, mine warfare, and nonlethal weapons. Prior to joining the Naval Studies Board, he was the lead mechanical engineer at Sensytech, Inc. (formerly S.T. Research Corporation), where he provided technical and program management support for satellite Earth station and small-satellite design. He received his Ph.D. in mechanical engineering from Vanderbilt University in 1995; his doctoral research was conducted at the Naval Research Laboratory (NRL), where he used an atomic force microscope to measure the nanomechanical properties of thin-film materials. In parallel with his graduate student duties, Dr. Draper was a mechanical engineer with Geo-Centers, Inc., working on-site at NRL on the development of an underwater X-ray backscattering tomography system used for the nondestructive evaluation of U.S. Navy sonar domes on surface ships.

Arul Mozhi is senior program officer at the National Research Council's Naval Studies Board and served as senior program officer at the NRC's Board on Manufacturing and Engineering Design and National Materials Advisory Board. Prior to joining the NRC in 1999, Dr. Mozhi was senior scientist and program manager at UTRON, Inc., a high-tech company in the Washington, D.C., area, working on pulsed electrical and chemical energy technologies applied to materials processing. From 1989 to 1996, Dr. Mozhi was a senior engineer and task leader at Roy F. Weston, Inc., a leading environmental consulting company working on long-term nuclear materials behavior and systems engineering related to nuclear waste transport, storage, and disposal in support of the U.S. Department of Energy. Before 1989 he was a materials scientist at Marko Materials, Inc., a high-tech firm in the Boston area, working on rapidly solidified materials. He received his M.S. and Ph.D. degrees (the latter in 1986) in materials engineering from the Ohio State University and then served as a postdoctoral research associate there. He received his B.S. in metallurgical engineering from the Indian Institute of Technology in 1982.

B

Workshop Agenda

J. ERIK JONSSON WOODS HOLE CENTER
WOODS HOLE, MASSACHUSETTS

Tuesday, July 27, 2004

Closed Session: Committee Members and NRC Staff Only

0800 CONVENE—Welcome, Composition and Balance Discussion
John D. Christie, Committee Chair
Charles F. Draper, Acting Director, Naval Studies Board (NSB)

Data-Gathering Meeting Not Open to the Public: Classified Discussion (Secret) Session Focus: Capabilities-Based Planning and Analysis Within the Department of Defense[1]

0830 WORKSHOP KICKOFF REMARKS (VIA TELECONFERENCE)—Need for Capabilities-Based Planning
Gregory Melcher, Deputy Director, Assessments Division, Office of the Deputy Chief of Naval Operations for Resources, Requirements, and Assessments, N81B

[1]Before the workshop, the Office of the Deputy Chief of Naval Operations for Manpower and Personnel (N1) briefing was provided to the committee by Richard Robbins, (N1Z), "N1 and Capabilities-Based Planning," at the Navy Annex, Washington, D.C., on July 21, 2004.

0900 JOINT DEFENSE CAPABILITIES: OSD (OFFICE OF THE SECRETARY OF DEFENSE) PERSPECTIVE
Vance Gordon, Program Analysis and Evaluation, Office of the Secretary of Defense

1045 JOINT DEFENSE CAPABILITIES: J8 (JOINT STAFF) PERSPECTIVE
LTC Robert F. Larsen, USA, Chief, Plans Branch, Joint Capability Division, Joint Staff, J8 JCD

1300 CAPABILITIES-BASED PLANNING AND ANALYSIS: NWDC (NAVY WARFARE DEVELOPMENT COMMAND) PERSPECTIVE
Wayne I. Perras, Director for Transformation, Navy Warfare Development Command

1445 CAPABILITIES-BASED PLANNING AND ANALYSIS: N6/N7 (DEPUTY CHIEF OF NAVAL OPERATIONS FOR WARFARE REQUIREMENTS AND PROGRAMS) PERSPECTIVE
CAPT Terry McKnight, USN, Warfare Integration Branch Head, Office of the Deputy Chief of Naval Operations for Warfare Requirements and Programs, N708

1615 CAPABILITIES-BASED PLANNING AND ANALYSIS: N8 (DEPUTY CHIEF OF NAVAL OPERATIONS FOR RESOURCES, REQUIREMENTS, AND ASSESSMENTS) PERSPECTIVE
LCDR Kenneth M. Masson, USN, Office of the Deputy Chief of Naval Operations for Resources, Requirements, and Assessments, N815F

Closed Session: Committee Members and NRC Staff Only

1745 COMMITTEE DISCUSSION: RECAP OF DAY 1
John D. Christie, Committee Chair

Wednesday, July 28, 2004

Closed Session: Committee Members and NRC Staff Only

0800 CONVENE—Committee Discussion, Plans for Day 2
John D. Christie, Committee Chair
Arul Mozhi, NSB Senior Program Officer

Data-Gathering Meeting Not Open to the Public: Classified Discussion (Secret) Session Focus: Capabilities-Based Planning and Analysis Within the Navy

0830 CAPABILITIES-BASED PLANNING AND ANALYSIS: N4 (DEPUTY CHIEF OF NAVAL OPERATIONS FOR FLEET READINESS AND LOGISTICS) PERSPECTIVE
Ariane L. Whittemore, Acting Deputy Chief of Naval Operations for Fleet Readiness and Logistics, N4 (via videoteleconference)

1015 ROLE TO IMPROVE CAPABILITIES-BASED PLANNING AND ANALYSIS: SYSCOM (SYSTEMS COMMANDS) PERSPECTIVE
RADM Kenneth D. Slaght, USN, Commander, Space and Naval Warfare Systems Command (via videoteleconference)
Patrick M. McLaughlin, Deputy Assistant Commander for Logistics and Aviation Depots, Naval Air Systems Command
Jim Egeland, Executive Director, Warfare Systems Engineering Directorate (NAVSEA 06B), Naval Sea Systems Command

Closed Session: Committee Members and NRC Staff Only
Session Focus: Report Deliberations, Administrative

1315 REPORT DISCUSSION AND CONSENSUS ON DRAFT CONCLUSIONS/RECOMMENDATIONS
COMMITTEE DISCUSSION: RECAP OF PRESENTATIONS AND QUESTIONS FOR ASSESSMENTS DIVISION, N81
John D. Christie, Committee Chair

Data-Gathering Meeting Not Open to the Public: Classified Discussion (Secret) Session Focus: Capabilities-Based Planning and Analysis Within the Navy

1500 DISCUSSION/Q&A WITH N81 LEADERSHIP (VIA VIDEOTELECONFERENCE)
RADM Joseph A. Sestak, Jr., USN, Director, Assessments Division, N81, and
Gregory Melcher, Deputy Director, Assessments Division, Office of the Deputy Chief of Naval Operations for Resources, Requirements, and Assessments, N81B

Closed Session: Committee Members and NRC Staff Only
Session Focus: Report Deliberations, Administrative

1600 REPORT DISCUSSION AND CONSENSUS ON DRAFT CONCLUSIONS/RECOMMENDATIONS

1730 END SESSION

Thursday, July 29, 2004

Closed Session: Committee Members and NRC Staff
Session Focus: Report Drafting

0830 Continue Report Discussion and Consensus on Draft Conclusions/Recommendations

1700 End Session

C

Acronyms and Abbreviations

AF/SA	Air Force Studies and Analyses
ASW	antisubmarine warfare
AVCS	Assistant Vice Chief of Staff
B-1	nuclear bomber (aircraft)
B-2	multirole bomber (aircraft)
C4ISR	command, control, communications, computers, intelligence, surveillance, and reconnaissance
CBP	capabilities-based planning
CBRNE	chemical, biological, radiological, nuclear, explosive
CJCS	Chairman, Joint Chiefs of Staff
CNO	Chief of Naval Operations
COCOM	combatant commander
CONOPS	concept of operations
CSG	carrier strike group
DCNO	Deputy Chief of Naval Operations
DOD	Department of Defense
ESG	expeditionary strike group
IRCA	Integrated Readiness Capability Assessment
JCIDS	Joint Capabilities Integration and Development System

JCS	Joint Chiefs of Staff
JROC	Joint Requirements Oversight Council
N1	Deputy Chief of Naval Operations (DCNO) for Manpower and Personnel
N4	DCNO for Fleet Readiness and Logistics
N6-/N7	DCNO for Warfare Requirements and Programs
N8	DCNO for Resources, Requirements, and Assessments
N70	Office of the DCNO for Warfare Requirements and Programs
N81	Assessments Division of the Office of the DCNO for Resources, Requirements, and Assessments
NAVAIR	Naval Air Systems Command
NAVSEA	Naval Sea Systems Command
NSS	Naval Simulation System
NWDC	Navy Warfare Development Command
O&M	operations and maintenance
OJCS	Office of the Joint Chiefs of Staff
OP-96	Systems Analysis Division of the Office of the Chief of Naval Operations (has been dissolved)
OPNAV	Office of the Chief of Naval Operations
OSD	Office of the Secretary of Defense
PA&E	Program Analysis and Evaluation
PNT	positioning, navigation, and timing
POM	Program Objective Memorandum
PPBE	planning, programming, budgeting, and execution
PPBES	planning, programming, budgeting, and execution system
PPBS	planning, programming, and budgeting system
QDR	Quadrennial Defense Review
R&D	research and development
RDT&E	research, development, testing, and evaluation
SECNAV	Secretary of the Navy
SPAWAR	Space and Naval Warfare Systems Command
SPG	Strategic Planning Guidance
SYSCOM	Systems Command
USD(AT&L)	Under Secretary of Defense (Acquisition, Technology, and Logistics)

REVIEW OF NASA'S Longitudinal Study OF Astronaut Health

David E. Longnecker, Frederick J. Manning,
and Melvin H. Worth, Jr., Editors

Committee on the Longitudinal Study of Astronaut Health

Board on Health Sciences Policy

INSTITUTE OF MEDICINE
OF THE NATIONAL ACADEMIES

THE NATIONAL ACADEMIES PRESS
Washington, D.C.
www.nap.edu

THE NATIONAL ACADEMIES PRESS • 500 Fifth Street, N.W. • Washington, DC 20001

NOTICE: The project that is the subject of this report was approved by the Governing Board of the National Research Council, whose members are drawn from the councils of the National Academy of Sciences, the National Academy of Engineering, and the Institute of Medicine. The members of the committee responsible for the report were chosen for their special competences and with regard for appropriate balance.

This study was supported by Contract/Grant No. NASW-03031 between the National Academy of Sciences and the National Aeronautics and Space Administration. Any opinions, findings, conclusions, or recommendations expressed in this publication are those of the Institute of Medicine Committee on the Longitudinal Survey of Astronaut Health and do not necessarily reflect the views of the organizations or agencies that provided support for the project.

International Standard Book Number 0-309-09148-9 (Book)
International Standard Book Number 0-309-53016-4 (PDF)

Additional copies of this report are available from the National Academies Press, 500 Fifth Street, N.W., Lockbox 285, Washington, DC 20055; (800) 624-6242 or (202) 334-3313 (in the Washington metropolitan area); http://www.nap.edu.

Printed in the United States of America.

The serpent has been a symbol of long life, healing, and knowledge among almost all cultures and religions since the beginning of recorded history. The serpent adopted as a logotype by the Institute of Medicine is a relief carving from ancient Greece, now held by the Staatliche Museen in Berlin.